DEPARTMENT OF ENGINEERING RESEARCH
UNIVERSITY OF MICHIGAN
ANN ARBOR

THE ELEMENTS
OF
METAL CUTTING

ORLAN W. BOSTON

Associate Professor of Shop Practice
and
Director of the Department of Engineering Shop
University of Michigan

T0350031

ENGINEERING RESEARCH BULLETIN
No. 5
December, 1926

Paperback ISBN : 978-0-472-75043-6

CONTENTS

ACKNOWLEDGMENTS

The present bulletin, which was presented at the Annual Meeting of the American Society of Mechanical Engineers in New York, December 6, 1926, forms a part of a general research program of the Society, and represents a contribution of the Machine Shop Practice Division and Research Sub-Committee on Cutting and Forming of Metals. It was made possible by the financial support secured through the Department of Engineering Research from the Michigan Manufacturers' Association, which through its Manufacturing Methods Committee sponsored the investigation. It represents, therefore, an investigation of mutual interest to the A.S.M.E., and the Department of Engineering Research at the University of Michigan.

The following Michigan manufacturers contributed their financial support to the project:

Detroit Steel Products Co., Detroit
Detroit Twist Drill Co., Detroit
Ford Motor Co., Detroit
General Motors Co., Detroit
Hupp Motor Co., Detroit
Michigan Screw Co., Lansing
National Twist Drill & Tool Co., Detroit
Packard Motor Car Co., Detroit
Reo Motor Car Co., Detroit
Russell Wheel & Foundry Co., Detroit
Timken Detroit Axle Co., Detroit
Detroit Copper & Brass Rolling Mills, Detroit
General Manufacturing Co., Detroit
Wilton Tool and Manufacturing Co., Detroit.

Valuable assistance was rendered in the preparation of equipment, in the execution of the experiments, and in the compilation of data, by Messrs. Carlton H. Currie, Donald L. Perkins, and Clarence J. Swigert.

SYNOPSIS

This paper gives an account of an investigation in the fundamental elements of metal cutting conducted in the Machine Tool Laboratory at the University of Michigan. The object of the investigation was to determine a relation between the force on the tool in the direction of cut for a constant cutting speed of 20 feet per minute, and the degrees of tool sharpness, the various tool angles, the width and depth of cut, and the physical properties of the materials cut. Nine representative types of material were cut including three carbon steels, three alloys steels, brass, and annealed and unannealed cast iron. The cutting was confined to straight-line motion on a planer, and the tools used were of the end-cutting type. No cutting fluids were used, and but one element was varied at a time.

The results show that the clearance angle has no influence on the force on the tool so long as the tool does not drag on the work; that the force on the tool remains constant for a wide variation of keenness of cutting edge and for thick chips, particularly, the tool edge may be rounded to 1/64 in. diameter without appreciable increase in the cutting force. It is also shown that the cutting force on the tool is reduced in direct proportion to the increase in front-rake angle, all other factors remaining constant. It is shown that thick chips are removed more efficiently than thin chips, and that narrow chips are removed more efficiently than wide chips. The results also indicate that there is an apparent relation between some of the physical properties of the metals and their machinability or the cutting force on the tool for the carbon steels in one group, the alloy steels in a second group, and cast iron in a third group.

DURING the last three years an investigation has been conducted as a project of the Department of Engineering Research in the Machine Tool Laboratory of the University of Michigan on elementary principles involved in the cutting of metals.

2 At the beginning, it was decided that the experiments should be confined to an investigation of a basic, scientific, rather than a practical, applied, nature. It was felt that the influence of curvature of the surface cut should be eliminated by confining the cuts to straight lines; also that each cut should be of sufficient length to permit the conditions of cutting to become uniform at the instant the reading was taken. In order to secure a straight-line cut having a constant cutting speed throughout the stroke, a planer was selected as the machine tool on which the work would be done.

3 The problems undertaken were as follows:

a To determine the influence of the degree of sharpness of the cutting edge of the tool, as the only variable, on the force on the tool or the energy required to remove a given volume of metal

b To determine the influence of the clearance back of the cutting edge of the tool, as the variable, on the force on the tool or the energy required to remove a given volume of metal

c To determine the influence of the front rake of the tool, as the variable, on the force on the tool or the energy required to remove a given volume of metal

d To determine the influence of side rake (skew), as the variable, on the force on the tool or the energy required to remove a given volume of metal

e To determine the influence of the depth of cut or width of cut, as the variable, on the force on the tool or the energy required to remove a given volume of metal

f To find a relation between the force on a tool of a given shape required to remove a specific chip of a given material and its physical or chemical properties.

THE TOOLS

4 All tools used in these tests were of the cut-off or end-cutting type. They were prepared for each problem in groups identical as to geometric form, heat-treatment, grinding, etc., except for the single variable under consideration. In some instances the group of tools for one problem was reforged or machined into tools of another form for another problem. They were all machined or forged from carbon or high-speed steel bars $2\frac{1}{8}$ by $1\frac{1}{4}$ in. cross-section and about 12 in. long. Fig. 1 is presented to illustrate the terms used for the parts and angles of the tools

and the size of the chip. At A is shown a front and side view of a front-rake tool. The front-rake angle is the angle between the tool face and the vertical plane A-A. F. W. Taylor called this angle the back slope. The clearance is the angle between the body of the tool back of the cutting edge, or flank, and the work. The cutting angle is the sum of the lip and clearance angles. At B is shown a tool which has both front and side rake. The cutting edge is in a horizontal plane. The side-rake angle is best shown in the plan view as that angle between the cutting edge BO and the vertical line BB. It may also be measured in any plane X-X perpendicular to the paper.

5 A typical record sheet for each piece of steel from which a tool was made is given in Appendix No. 1-A. The first tool was designated as A-1, its chemical composition, whether machined

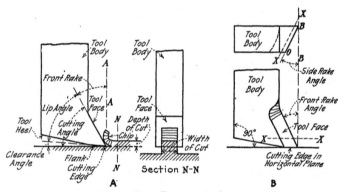

FIG. 1 TOOL PARTS AND ANGLES

or forged, heat-treatment, angles, hardness, etc., given. When this tool was modified in any way, it was subsequently designated as another tool, A-2, etc. Every experiment sheet shows the tool used in that test. A table listing all tools and indicating the problems for which each was used is given in Appendix No. 1-B.

6 The tools were hardened and drawn in accordance with the specifications received from the manufacturer of the steel. Endurance was not an essential feature of the test. It was, however, necessary that a tool retain its geometric form with very little variation throughout a test. Usually the test was not of sufficient length to cause failure of the tool. Some of the tools were hardened in a forge fire and others in gas-fired furnaces, the temperatures of which were definitely determined by thermocouples. Our results show that the furnace heat-treated tools give more consistent results than those treated in the forge fire. This is confirmed by tests conducted at the Bureau of Standards as described in their technical news bulletin No. 105 of January, 1925. It was

found that comparable performance was obtained when tools were raised to approximately equal temperatures for equal times in hardening. The Bureau test failed to show the superiority of the forge-fire hardening over gas, oil-fired, or electric furnace hardening.

METHOD OF MEASURING FORCES

7   Fig. 2 shows the Liberty 30-in. by 36-in. by 8-ft. planer on which the tests were conducted. The power was furnished from

Fig. 2   PLANER AND DYNAMOMETER USED IN EXPERIMENTS

a main drive shaft, a satisfactory arrangement in that the horse-power output of the machine was measured rather than the input. The cutting tool is shown mounted in the standard head on the cross-rail. The specially designed dynamometer is shown mounted on the planer table. The material being cut is clamped on the bed of the dynamometer.

8 Fig. 3 shows a line diagram of the dynamometer which consists essentially of a cast-iron bed B weighing about 1300 lb., mounted on seven horizontal knife edges C. The table is prevented from moving sideways by the four horizontal floating pins shown in the plan view which are supported by the brackets F. The bed is prevented from being lifted by four knife edges shown at G. An initial load of 2000 lb. tending to force the bed to the left was put on the bed by compressing the loop spring D. This loop spring was made of ¾-in. square spring steel having major and minor axes to center of bar of 11½ and 4¾ in., respectively, and had previously been calibrated on a Riehlé testing machine. It

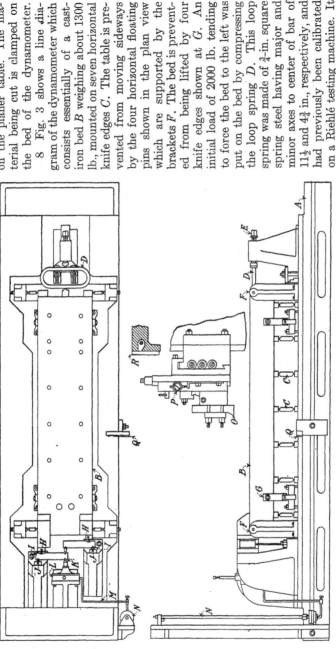

FIG. 3 LINE DIAGRAM OF DYNAMOMETER

was found to give more satisfactory results than helical springs, as identical load-strain curves were obtained for increasing or reducing increments of load. Fig. 4 shows a close view of the

FIG. 4   PLANER TOOL HEAD AND LOOP SPRING
(Also shows indicator for measuring depth of cut.)

loop spring in its position between the tail casting and the dynamometer bed. Fig. 5 shows the calibration curve up to 7000 lb. and gives consistent values of 0.0314 in. deflection per 1000 lb. of load. The deflections were read on a dial gage mounted across the minor axis. The 2000-lb. initial load was transmitted by the vertical knife edges *H* through the levers *I*, about the fulcrum *J*, to the pin *K*, which was mounted on a piston supported so as to bear against a rubber diaphragm contained in the cylinder *L*. This pressure was then transmitted hydraulically through the tube *M* to the mercury-column gage *N*.

9   Fig. 6 shows a separate view of the differential mercury-column gage. The point of zero reading on the mercury column corresponded with the initial spring load of 2000 lb. on the dynamometer bed. The mercury column was next calibrated to read in pounds the actual horizontal load in the direction of the

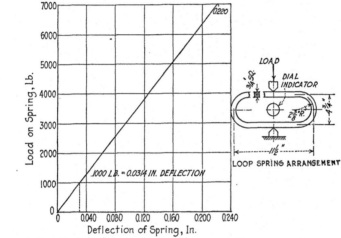

FIG. 5   LOOP-SPRING ARRANGEMENT AND CALIBRATION CURVE

tool travel, by means of the loop spring *D*. Originally 100- and 200-lb. test gages were used but they gave unsatisfactory results.

10   The scale on the mercury column was made up by marking the mercury level for each successive load on the dynamometer bed as obtained by compressing the loop spring a definite amount with the adjustable screw *E*, Fig. 3. A 1000-lb. load of the spring produced a 2.539-in. rise of mercury in the column. While the dynamometer was designed for a load of 14,000 lb., it was calibrated only to 7000 lb.

11   In order that all readings might be taken in the same position of the bed, a spring contact was prepared, half·of which was mounted on the vertical column as shown at *R* in the plan view of Fig. 3, the other half mounted on the bed of the planer shown at *Q*. As these springs passed, there was a snap, at which time the reading of the mercury column was taken. Originally all cuts

were made 4 ft. in length. It was found, however, that consistent readings could be obtained with a length of cut of 30 to 36 in.

12   The calibration was made with only the dead weights of the dynamometer parts plus a 300-lb. bar of material acting as vertical force. However, it is assumed that due to the ample support applied by the seven knife edges under the dynamometer table and

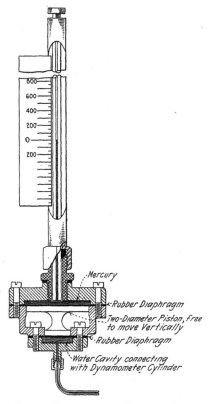

Fig. 6   Differential Column Gage

their very slight motion, any additional vertical loads caused by tool action would not cause sufficient friction in these knife edges to influence the force reading in a horizontal direction. This calibration demonstrated that a given increase in load applied to the dynamometer table caused a given rise in the mercury column regardless of what the total load acting might be, which proved that whatever friction existed did not influence the action of the dynamometer.

13 The dynamometer is sensitive to a force of ten pounds which may be read on the mercury-column gage. The variation of the actual chip size from the desired, accounts for some variations in readings as well as the variation caused by the material itself. Each test, however, was an average of at least six consecutive readings so that the low force of a light chip would be compensated by the higher force of the heavier chip which would next be removed.

THE MATERIAL

14 The materials selected to be cut in the experiments were confined to those in most common use which would give a wide range of physical characteristics. They included straight carbon steels (low, medium, and high carbon), $3\frac{1}{2}$ per cent nickel steels of high and low carbon, nickel-chromium steel, cast iron, and brass.

15 The steel was furnished in bars 4 by 6 in. and 48 in. long, prepared for the experiments by milling parallel grooves $\frac{3}{8}$ to $\frac{1}{2}$ in. wide and 1 in. deep in the upper surface, lengthwise of the bar. The material left between these grooves formed lands, the widths of which were equal to the width of chip to be cut. Fig. 2 shows one of these bars clamped on the bed of the dynamometer ready for testing. The tool was set over one land as shown in Figs. 2 and 4, and each successive cut was taken by feeding the tool vertically downward. The feed of the tool was accurately measured on the dial gage mounted on the tool holder as shown in Fig. 4.

16 The cast-iron bars were all cast from one ladle in the foundry laboratory in order to get bars as nearly alike as possible. Six bars were cast in a mold, each bar being $1\frac{3}{4}$ in. wide by 4 in. deep and 48 in. long. The lands were prepared on the top of the bar by milling.

17 Brass was furnished in rolled sheets, half-hard, of various thicknesses so that when cut on the edge, the thickness of the sheet represented the width of the chip. The 1-in. thick brass sheet was milled on its upper edge so as to present lands of desired widths. This also removed the cold-rolled skin from the side of the land.

18 A material record sheet for each bar of material was kept, which gave all information as to manufacturer, chemical analysis, heat-treatment, hardness, and other physical properties. Appendix No. 2-A is a copy of the sheet for bar No. 1 of machine steel. Appendix No. 2-B is a material-record-sheet summary showing those problems for which each bar was used. Appendix No. 3 gives a list of test bars used for obtaining the physical properties of the materials cut.

19 Table 1 shows a list of the material, giving its physical and chemical properties. Table 2 is a summary sheet showing the cutting force for various tools for each material. These tables are referred to later in connection with individual problems.

TABLE 1 PHYSICAL AND CHEMICAL PROPERTIES OF MATERIALS CUT

Physical strengths given in lb. per sq. in.

Material	Bar No.	Tension					Compression			Shear	Hardness		
		Elastic limit	Ultimate strength¹	Reduction of area, per cent	Elongation, per cent 2 in. g.l.	Type of failure	Elastic limit	Ultimate strength¹	Type of failure	Ultimate strength¹	Brinell	Scleroscope	Rockwell
S.A.E. 3120	31	47,500	77,000	70.2	30.3	cup and cone	41,700	116,750	......	54,150	155	24	48
S.A.E. 2345	29	60,000	99,670	42.7	25	cup and cone and plane ⊥ to axis	51,550	140,500	......	61,650	191	26	59
S.A.E. 2320	30	55,250	79,500	55.6	30	cup and cone and plane ⊥ to axis and 45° shear	42,750	119,000	......	54,400	162	24	55
S.A.E. 1035	28	19,900	54,000	60	35	....	16,000	93,000	......	37,700	99	19	48
0.15% C	1	25,300	52,400	67	41	cup and cone	22,000	80,500	......	39,400	106	21	54
0.15% C	2	24,500	52,000	66	40	cup and cone	......	......	......	39,400	102	22	56
0.15% C	3	23,000	51,300	67	40	cup and cone	......	......	......	39,400	101	21	58
0.15% C	4	20,900	52,000	65	40	cup and cone	......	......	......	39,410	102	21	45
0.15% C	5	22,400	50,300	63	42	cup and cone	19,075	85,350	......	39,400	101	26	56
0.15% C	6	22,500	52,750	63	40	cup and cone	......	......	......	39,400	99	21	56
1.08% C	32	39,170	80,000	56.7	32	cup and cone and plane ⊥ to axis	42,000	112,000	......	52,670	155	25	52
C.I.	1	none	16,160	none	none	plane ⊥ to axis	none	62,600	pl. 40° to axis	26,960	137	28	70
C.I.	2	none	18,200	none	none	plane ⊥ to axis	none	69,300	pl. 33° to axis	26,000	156	28	83
C.I.	3	none	17,200	none	none	plane ⊥ to axis	none	70,000	pl. 34° to axis	24,000	146	30	76
C.I.	3A	none	14,000	none	none	....	none	57,200	pl. 34° to axis	24,700	118	25	65
C.I.	4	none	16,800	none	none	......	none	72,000	pl. 25° to axis	23,800	143	30	76
C.I.	4A	none	14,600	none	none	plane ⊥ to axis	none	59,100	pl. 25° to axis	20,500	143	22	64
C.I.	5	none	17,300	none	none	......	none	64,900	pl. 32° to axis	27,000	153	31	78
C.I.	6	none	16,700	none	none	plane ⊥ to axis	none	65,700	pl. 32° to axis	28,400	154	32	....
C.I.	7A	none	14,520	none	none	plane ⊥ to axis	none	59,780	pl. 37° to axis	21,200	118	31	65
C.I.	8	none	17,200	none	none	plane ⊥ to axis	none	73,000	pl. 33° to axis	22,700	149	22	85
C.I.	8A	none	16,900	none	none	plane ⊥ to axis	none	68,600	pl. 34° to axis	19,100	109	32	57
C.I.	9	none	21,300	none	none	plane ⊥ to axis	none	80,300	pl. 27° to axis	28,300	179	24	88
C.I.	9A	none	13,400	none	none	plane ⊥ to axis	none	60,200	pl. 35° to axis	24,400	121		62
C.I.	10	none	......	none	none	......	none	......	......	25,000	......		....
C.I.	11	none	18,900	none	none	plane ⊥ to axis	none	77,000	pl. 33° to axis	......	121	26	75
Brass	24	....	58,000	43	28	plane 45°	......	......		35,600	124	24	70
Brass	25	....	60,700	48	34	plane 45°	......	......		34,600	116	20	65
Brass	26	35,000	58,000	32	31	plane 45°	......	......		35,600	124	24	70
Brass	27	49,700	58,548	51	...	plane 45°	......	......		35,903	121	21	72
Brass	33	25,600	44,877	36	...	plane 45°	......	......		28,360	107	17	50
Brass	34	57,300	59,831	29		plane 45°	......	......		36,060	131	24	71

¹ Ultimate strength = maximum load/original area. Brinell Nos. from 3000 kg. load on 10-mm. ball. Rockwell nos. from 1/16-in. diam. ball.

CHEMICAL ANALYSES

S.A.E. 3120 steel, bar No. 31: C, 0.17; Mn, 0.60; P, 0.014; S, 0.019; Cr, 0.67;
S.A.E. 2345 steel, bar No. 29: C, 0.48; Mn, 0.59; P, 0.010; S, 0.024; Ni, 3.47; Si, 0.31; Ni, 1.28.
S.A.E. 2320 steel, bar No. 30: C, 0.20; Mn, 0.65; P, 0.014; S, 0.018; Ni, 3.87; Si, 0.18.
S.A.E. 1035 steel, bar No. 28: C, 0.21; Mn, 0.27; P, 0.022; S, 0.080; Si, none.
0.15 per cent carbon steel, bars Nos. 1 to 6 incl.: C, 0.15; Mn, 0.24; Si, 0.13; S, 0.029; P, 0.014.
1.03 per cent carbon steel, bar No. 32: C, 1.03; Mn, 0.28; P, 0.017; S, 0.018; Si, 0.18.
Cast iron, all bars: C, 3.52; Si, 2.18; S, 0.108; Mn, 0.69; P, 0.489.
Brass, bars Nos. 24 and 26: Cu, 61.34; Pb, 1.44; Fe, 0.034; Zn, r emainder.
Brass, bar No. 25: Cu, 61.48; Pb, 1.615; Fe, 0.069; Zn, remainder.
Brass, bar No. 27: Cu, 60.68; Pb, 1.635; Fe, 0.132; Zn, remainder.
Brass, bar No. 33: Cu, 61.98; Pb, 1.49; Fe, 0.050; Zn, 36.48.
Brass, bar No. 34: Cu, 64.10; Pb, 1.52; Fe, 0.042; Zn, 34.34.

TABLE 2 CUTTING FORCES

Pounds per 0.001 sq. in. of cross-sectional area of cut

Material	Bar No.	Clearance in degrees, 30° front rake, 0° side rake, ½″ width of cut, 0.024″ depth of cut					Front-rake angle in degrees, 4° clearance, 0° side rake, ½″ width, 0.012″ depth of cut					Side-rake angle in degrees, 30° front rake, 4° clear., ½″ width, 0.012″ depth of cut				Width of cut, 30°-0°-4° tool, 0.012″ depth of cut				Depth of cut in inches, 30°-0°-4° tool, ½″ width					
		2	4	6	8	10	0	10	20	30	40	0	10	20	30	⅛″	¼″	½″	¾″	.003	.006	.012	.024	.032	.048
S.A.E. 3120	31	....	....	....	....	....	none	294	255	218	180	218	216	222	226	197	209	218	....	331	265	218	192	184*	172
S.A.E. 2345	29	....	....	....	....	....	none	314	293	275	247	255	265	—	—	225	246	275	....	385	307	275	214	203	....
S.A.E. 2320	30	....	....	....	....	....	none	295	260	229	195	229	241	244	254	195	210	229	....	348	276	229	191	184	....
S.A.E. 1035	28	....	....	....	....	....	none	296	245	214	172	214	—	220	216	210	....	214	217	—	253	214	180	....	....
0.15% C	1	....	....	....	207	219	....	....	....	....	....	246*	....	....	....	231*	244	246*	....	....	288*	246*	212*	199	....
0.15% C	2	....	....	....	....	....	....	....	....	....	....	....	....	....	....	....	220	....	....	....	....	none	....	....	....
0.15% C	3	....	....	206	....	....	....	332	275	233	191	233	....	....	....	....	222*	233	....	....	283	233	208	....	....
0.15% C	4	209	207	....	....	....	....	307	—	234*	....	234*	....	....	....	206	—	234*	....	349	279	234	207	....	....
0.15% C	5	....	....	....	....	....	....	....	....	237*	....	237*	....	....	....	217	226	237*	252	....	292	237*	207	....	....
0.15% C	6	....	....	....	....	....	....	....	....	246	....	246	....	....	246	....	230	246	....	....	307	246	212	....	....
1.03% C	32	....	....	....	....	....	....	414	364	305	262	313	305	314	297	271	285	305	....	444	368	313	256	234	....
C. I.	1	....	....	....	....	....	....	....	....	....	....	....	....	....	....	....	....	....	....	....	125	98	....	....	....
C. I.	2	....	....	....	....	....	....	....	....	122	....	122	131	128	120	....	....	122	....	....	....	122	....	....	....
C. I.	3	....	....	....	....	....	....	....	....	....	....	....	....	....	....	....	....	....	....	....	....	none	83	....	....
C. I.	3A	....	....	....	....	....	....	....	....	....	....	....	....	....	....	....	....	....	....	....	....	none	72	....	....
C. I.	4	86	84	91	89	90	....	....	....	....	....	....	....	....	....	....	....	....	....	....	135	none	84	....	....
C. I.	4A	....	....	....	....	....	....	....	....	91*	....	91*	....	....	....	....	....	91*	....	....	115	91*	73	....	59
C. I.	5	....	....	....	....	....	....	164	129*	111	99.6	111	....	....	....	....	....	111	....	....	141	111	....	....	....
C. I.	6	....	....	....	....	....	....	....	....	....	....	....	....	....	....	....	....	....	....	....	125	none	82	....	....
C. I.	7A	....	....	....	....	....	129	....	....	86	....	86	....	....	....	....	91	86	94	....	....	86	....	....	....
		10° front rake tool																							
C. I.	8	125	125	125	130	127	....	....	....	103*	....	103*	....	....	....	....	....	103*	....	....	127	103*	82	....	65
C. I.	8A	....	....	....	....	....	....	....	....	90*	....	90*	....	....	....	....	....	90*	....	....	108	90*	74	....	60
C. I.	9	....	....	....	....	....	152	134	112	95	86	95	....	....	....	....	....	95	....	....	125	95	....	....	....
C. I.	9A	....	....	....	....	....	127	116	101	88	75	88	....	....	....	....	....	88	....	....	106	88	....	....	....
												0° front rake													
C. I.	10	....	....	....	....	....	143	....	....	....	....	143	145	142	147	....	....	....	....	....	....	....	....	....	....
C. I.	11	....	....	....	....	....	171	....	....	102	....	102	....	....	....	....	97.5	102	103	....	....	102	....	....	....
		0° front rake			0.012 in. deep																				
Brass	24	127	126	126	....	126	126	109	92	79	....	79	....	....	....	....	....	79	....	....	86	79	74	....	....
Brass	25	....	....	....	....	....	....	....	....	....	....	....	....	....	....	....	....	....	....	....	81	76	74	....	....
Brass	26	....	....	....	....	....	125	....	....	74	....	74	74	82	82.5	....	....	74	....	....	....	74	....	....	....
Brass	27	....	....	....	....	....	....	....	....	....	....	....	....	....	....	....	....	....	....	....	....	....	....	....	....
Brass	33	....	....	....	....	....	....	....	....	....	....	....	....	....	....	....	....	....	....	....	....	....	....	....	....
Brass	34	....	....	....	....	....	....	....	....	....	....	....	....	....	....	....	....	....	....	....	....	....	....	....	....

* Value taken from curve. Available data not reliable. All cutting speeds, 20 ft. per min.

The Tests and Results

20   For clarity each problem is treated below individually.

Sharpness of Tool — Problem (a)

21   Problem (a) is " the determination of the influence of the degree of sharpness of the cutting edge of the tool, as the only variable, on the force on the tool or the energy required to remove a given volume of metal." This problem may also be to determine a standard practical degree of tool sharpness. It is divided into two parts. It was necessary to make sure that the tools as ordinarily heat-treated and ground would present a cutting edge which would give uniform results throughout any given test. For this purpose, tools having 0-, 10-, 20-, 30-, and 40-deg. front-rake angles with a clearance angle of 6 deg. were prepared, hardened, and drawn in accordance with the specifications of the manufacturer of the tool steel, some in a forge fire and others in a furnace. Depths of cuts of 0.005, 0.020, and 0.040 in. all having a width of 0.500 in. were made in open-hearth 0.15 per cent carbon steel fully annealed. The cutting speed was kept constant at 20 ft. per min.

22   The procedure was to grind the tool and cut with it under uniform conditions until the edge became dull and to observe the length of life of the cutting-tool edge which would make no increase in the reading of the force on the tool. It was found that in order to secure a cutting edge which would remain practically constant when cutting between 400 to 1600 linear feet that the degree of hardness was critical, that is, the temperatures of hardening and drawing must be very carefully controlled. It was also found that a properly hardened tool of either carbon or high-speed steel will present a cutting edge which will take all the cuts necessary for any of the following tests without dulling sufficiently to affect the gage reading.

23   Visual examination of the cutting edge was resorted to during most of these tests as the wear or change of condition of the edge was so small and affected the force on the tool so little that microscopic examinations were not considered necessary. It appeared that the original keenness of the ground edge, as is felt by drawing the finger along it, disappeared after the first two or three feet of cutting. The tool then seemed to remain in this new condition while cutting several hundred feet. It was found that when the cutting edge again changed, it wore rapidly and the failure of the tool soon followed. As we were interested in the life of the tool while sharp rather than the actual time and nature of failure, detailed accounts of the failure are not recorded.

24   In several instances, tools were prepared with definite degrees of tool sharpness, that is, some were carefully honed and

others ground on a tool grinder equipped with fine and coarse wheels. Photomicrographs were made of the cutting edges, looking down on the face of the tool over which the chip would slide. Cuts of $\frac{1}{4}$ or $\frac{1}{2}$ in. width and 1 ft., 3 ft., 10 ft., etc., of various depths were taken, after which photomicrographs were again taken. A comparison of these graphs showed no indications of wear on the tool for these short cuts, nor could it be detected just what part of the cutting edge, which was longer than the chip was wide, did the cutting.

25   The second part of the test was to determine the influence of a definite condition of the cutting edge on the force as registered by the dynamometer. The tools used for this test are shown in Fig. 7. Each one has a clearance angle of 4 deg. but front

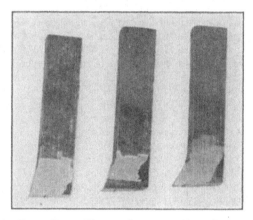

Fig. 7   Three Tools Used in Sharpness Test, Problem (a)
10-, 20-, and 30-deg. front rake angles, 4-deg. clearance.

rake angles of 10, 20, and 30 deg. respectively. They were hardened at 1500 deg. fahr. in water and tempered at 440 deg. fahr. in oil. Brinell and Rockwell hardness numbers taken about the cutting edges were consistent at 652 and 60-62, respectively, for all tools. The cutting speed was 20 ft. per min. in all cases. The tools were first carefully ground on a machine and then the cutting edges were honed so that under a microscope they appeared to be as even as the average razor blade. With this cutting edge, a series of cuts were taken in open-hearth 0.15 per cent carbon steel, annealed, and cast iron, with chips $\frac{1}{2}$ in. in width and 0.006 or 0.031 in. in depth. The force on the tool as indicated by the dynamometer was recorded. This force for comparative purposes was reduced to pounds per 0.001 sq. in. of cross-sectional area of the chip and plotted as ordinates over the condition of the cutting edge of the tool as abscissas. The tools were next ground in a tool grinder equipped with an alundum wheel of 60 grain and a similar set of readings recorded. Similar readings were subsequently taken with the tools having their cutting edges prepared successively by grinding with an alundum No. 36 grit wheel, ground deliberately blunt $\frac{1}{64}$ of an inch so as to give zero rake at the cutting edge, honed to an edge having $\frac{1}{256}$ in. diameter roundness, to $\frac{1}{128}$ in. diameter roundness, and finally to $\frac{1}{64}$ in. diameter roundness. These edges were carefully prepared and measured under a large magnifying glass, a steel rule being held close to the cutting edge.

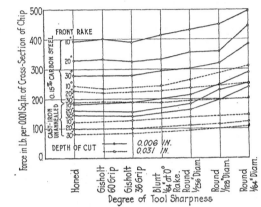

FIG. 8   UNIT FORCE-TOOL SHARPNESS CURVES FOR CAST IRON AND STEEL
PROBLEM *(a)*

Materials: Cast iron, unannealed, bars Nos. 7 and 10 0.15 per cent carbon steel, annealed, bars Nos. 3 and 5
Tools: Front rake, 10-, 20-, and 30-deg. clearance 4 deg.
Widths of cut: 0.5 and 0.25 in.
Depths of cut: 0.006 and 0.031 in.
Cutting speed: 20 ft. per min.

26   The results of the tool-sharpness tests are shown plotted in Fig. 8. The results for both cast iron and low-carbon steel are given for the three tools for each condition of the cutting edge. The solid lines represent the unit forces for the thin chips (0.006 in.). It is seen that the curves remain practically horizontal until the cutting edge is deliberately rounded to $\frac{1}{256}$ in. diameter when there is a slight increase. Beyond this point the increase in the unit force is more noticeable. For a degree of roundness of the cutting edge of $\frac{1}{128}$ in. diameter the increase does not seem to be as much as would be expected. While the values for cast iron are much lower on the scale than those for steel, there is a noticeable similarity in the results. The dashed lines are for the heavy chips (0.031 in. deep) and with the exception of the rounded edge of $\frac{1}{64}$ in. diameter of the 10-deg. front-rake tool cutting steel, there appears to be no appreciable increase in the force on the tool. It was concluded from these curves that for a tool carefully heat-treated, and ground on any standard commercial abrasive wheel,

there is no need to introduce a correction factor for the force on the tool for various degrees of sharpness or for reasonable wear during the test.

27 This series of tests does not indicate what happens to the force on the tool after the tool becomes dull by cutting.

28 Several references are made in the following pages to chip shape or other characteristics as it is thought that they may contribute some information of value to the subject. Fig. 9 has been prepared to show certain features of chip formation. It was observed that for a given depth of cut, the number of cross-lines on the steel land after a chip had been removed varied with the degree of tool sharpness. With the keen cutting edge there were

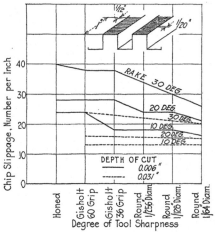

FIG. 9 CHIP SLIPPAGE — TOOL SHARPNESS CURVES

Material: 0.15 per cent carbon steel, annealed.
Tools: 4-deg. clearance.

more slips per inch which gave a better finish than with the dulled edge. The highest curve shows that for a 30-deg. front-rake tool having a depth of cut of 0.006 in. there are 40 slips per inch, which is reduced to 26 slips per inch for the dullest edge. There appears to be little difference in slippage for the first three conditions of tool sharpness; also the slippage variation is reduced for the heavier cuts as shown by the dashed lines. These lines may be due to the slippage of the metal over the tool in the formation of a chip or may be a function of chatter, although there was no obvious indication of chatter.

29 When cutting cast iron and for a given depth of cut, it was observed that there was no appreciable difference in surface condition after the removal of the chip by tools having various rake angles and degrees of tool sharpness. There is a very noticeable

chip slippage for the 0.031-in. depth of cut in cast iron, but the surface is so smooth that the number of slips per inch cannot readily be seen nor can they be accurately counted. There seem to be approximately 20 slips per inch for honed tools and 28 slips per inch for the $\frac{1}{64}$-in. rounded-edge tools. As with steel, it appeared in every case that the initial degree of tool sharpness was changed by the first cut of 3 ft. but that this new condition was then maintained for the remainder of the test.

30   It was noticed that the beginning of each steel shaving had a smooth, bright surface on the underside for about $\frac{1}{8}$ to $\frac{1}{2}$ in.,

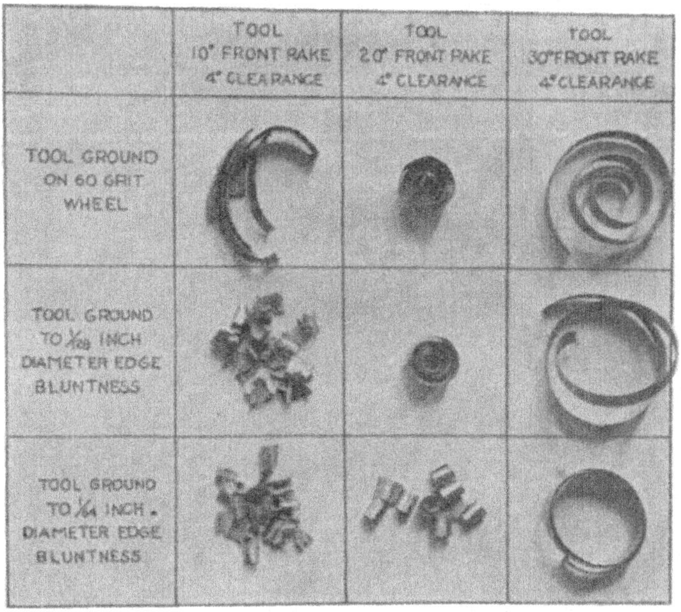

FIG. 10   TOOL SHARPNESS — FRONT RAKE, STEEL CHIPS

Material: 0.15 per cent carbon steel, annealed.
All chips 0.502 in. wide and 0.006 in. deep.

being longer for the greater thicknesses of chip, after which the rough slippage lines appeared and continued to the end of the cut. This seems to confirm the belief that after cutting has started the function of the cutting edge, particularly for the thicker chips, is to support a wedge of the material being cut. When the chip was broken off at the end of the cut, the small amount of material wedged on the cutting edge of the tool usually clung to the chips and was distinctly noticeable. The action of the cutting tool is not apparent to the eye; however an examination of the surface of the tool shows that there is a rubbing action a slight distance in from the edge, the distance varying with the thickness of the chip.

31 Fig. 10 shows some steel chips removed by the 10-, 20-, and 30-deg. front-rake tools each having successively the three cutting edges, ground on 60-grit wheel, honed to $\frac{1}{128}$-in. diameter edge, and honed to $\frac{1}{64}$-in. diameter edge. For the 10-deg. front-rake tool, the chips appear to become more broken with the increase in bluntness. This is not true with the 30-deg. front-rake tool. It is also observed from this figure that the increased front rake for each degree of bluntness gives a more continuous type of chip. Fig. 11 shows a similar display of cast-iron chips. This, too, shows

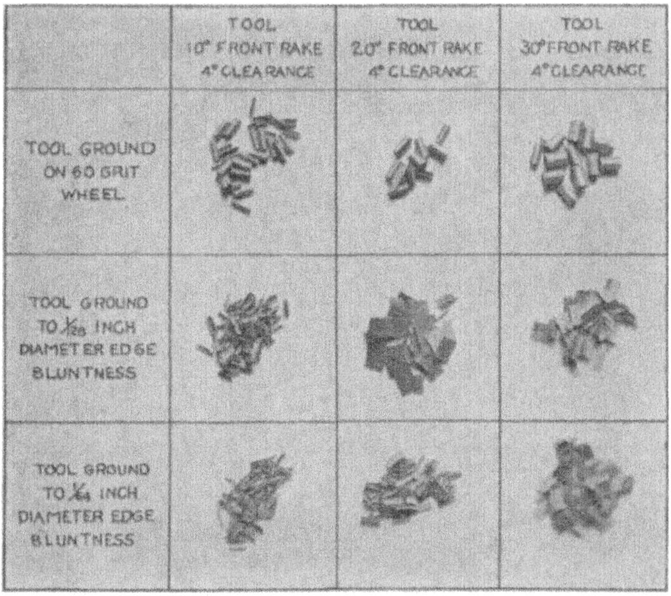

	TOOL 10° FRONT RAKE 4° CLEARANCE	TOOL 20° FRONT RAKE 4° CLEARANCE	TOOL 30° FRONT RAKE 4° CLEARANCE
TOOL GROUND ON 60 GRIT WHEEL			
TOOL GROUND TO $\frac{1}{128}$ INCH DIAMETER EDGE BLUNTNESS			
TOOL GROUND TO $\frac{1}{64}$ INCH DIAMETER EDGE BLUNTNESS			

Fig. 11 Tool Sharpness — Front Rake, Cast-Iron Chips

Material: Cast iron. All chips 0.502 in. wide and 0.006 in. deep.

the chips to be more broken up as the cutting edge is increased in bluntness. For the well-ground tool (60-grit wheel) the chips appear to be more continuous with the increased front rake. The chips under the 10-deg. front-rake tool vary from $\frac{1}{8}$ to $\frac{3}{4}$ in. in length while those for the 30-deg. front-rake tool are for the full length of cut, 30 in.

Tool Clearance — Problem (b)

32 Problem (b) is " to determine the influence of the clearance back of the cutting edge of the tool, as the variable, on the force on the tool or the energy required to remove a given volume of metal." A number of tools were prepared having 0-, 10-, 20-, 30-

and 40-deg. front-rake angles, respectively, each in turn ground
with clearance angles of 0, 2, 4, 6, 8, and 10 deg. The material
cut for this problem was confined at first to the low-carbon steel
fully annealed so that the results obtained would be based on a
material uniform throughout. The width of the cut in all cases
was 0.500 in., the depth of cut was successively 0.003, 0.006, 0.012,
and 0.024 in. The cutting speed in all cases was constant at 20

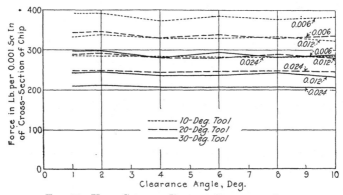

FIG. 12   UNIT CUTTING FORCE — CLEARANCE ANGLE

Material: Open-hearth 0.15 per cent carbon steel, annealed.
Tools: 10-, 20-, and 30-deg. front rake.
Speed: 20 ft. per min.

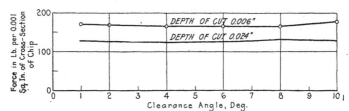

FIG. 13   CUTTING FORCE — CLEARANCE ANGLE, EXPERIMENT 19

Material: Cast iron, bar No. 8. Tool: 10-deg. rake.
Width of cut: 0.50 in. Speed: 20 ft. per min.

ft. per min. For comparative purposes, the force registered by
the dynamometer was reduced to pounds per 0.001 sq. in. of
cross-sectional area of the chip.

33 Representative values for the 10-, 20-, and 30-deg. front-
rake tools cutting 0.006, 0.012, and 0.024 in. depth of chip for
the low-carbon steel are shown in Fig. 12. While these curves are
drawn through the actual values obtained, it appears obvious that
their general tendency is to be straight horizontal lines. Results
for the 0-deg. clearance for the three tools on low-carbon steel are
not indicated because of their doubtful value, readings being alter-

nately high and low with an indication of considerable friction between the work and the clearance face of the tool. Also, values for a chip of 0.003 in. depth of cut are not reproduced, as the values for such thin chips appear to be too erratic to be of reliable assistance.

34 Fig. 13 shows unit-force values when cutting cast iron with a tool having a 10-deg. front rake angle for chips 0.006 and 0.024 in. deep and 0.500 in. in width. Fig. 14 shows the unit-force values for similar chips of cast iron for a tool having 30-deg. front rake.

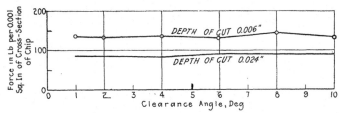

FIG. 14 CUTTING FORCE — CLEARANCE ANGLE, EXPERIMENT 19

Material: Cast iron, bar No. 4. Tool: 30-deg. rake.
Width of cut: 0.50 in. Speed: 20 ft. per min.

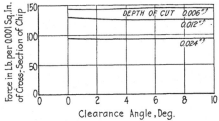

FIG. 15 CUTTING FORCE — CLEARANCE ANGLE

Material: Brass, bar No. 24. Tools: 0-deg. rake.
Width of cut: 0.506 in. Speed: 20 ft. per min.

The unit forces for the 0.006- and 0.024-in. depths of cuts are 165 and 125 lb., respectively, for the 10-deg. front-rake tool, and 132 and 85 lb., respectively, for the 30-deg. tool.

35 Fig. 15 shows representative values for brass, the tool used having 0-deg. front rake and depths of cut of 0.006, 0.012, and 0.024 in. Here again are found straight horizontal lines. Unit forces of 144, 126, and 99.3 are representative values for each depth of cut. The 0.024-in. depth of cut caused chatter of the tool for all clearance angles. The chip slippage marks on the land after the 0.006- and 0.012-in. depths of cut chips had been removed were not countable as the surface was so smooth. After the 0.024-in. chip was removed, slippage or chatter lines were evident

and between 22 and 28 per linear inch. The clearance angle had little effect on the length of chip removed. A full-length (3-ft.) chip, in form of a coil, was removed with 0.006-in. depth of cut, about two inches in length for the 0.012-in., and $\frac{1}{32}$ in. for the 0.024-in. Fig. 16 is a summary sheet showing the unit forces plotted against clearance angles for steel, cast iron, and brass with tools having 10-, 10-, and 0-deg. front rake respectively. The depth of cut in each case is 0.024 in. The values given in Table 2, unless noted to the contrary, are for a 30-deg. front-rake tool. The curves give more values than the table because of the many combinations of rake and depths of cut.

36   Conclusions may be drawn from the above representative data to the effect that a variation of the clearance angle has no

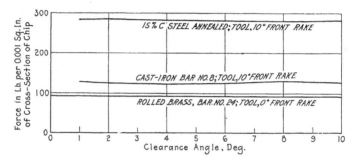

FIG. 16   SUMMARY UNIT CUTTING FORCE — CLEARANCE ANGLE CURVES, PROBLEM (b)

Materials: 0.15 per cent carbon steel, cast iron, and brass.
Depth of cuts: 0.024 in. Width of cuts: 0.5 in.
Speed: 20 ft. per min.

influence on the force required to remove the chip. Abrasion on the flank of the tool would probably prove excessive for low clearance angles, yet the less the clearance angle, the more the cutting edge is supported and the more metal there is to carry away the heat generated. It seems desirable, therefore, to select a clearance angle which will positively prevent the rubbing of the tool over the work and yet be as small as possible. In commercial practice, the accuracy with which these angles can be ground is one limiting feature, the feed of the tool is another. F. W. Taylor[1] recommended between 4- and 12-deg. clearance depending on the feed and the doubtful accuracy of grinding. As the feed, such as that given a side-cutting tool, does not enter into the work of these tests and as all tools are machine ground, a clearance angle of 4 deg. was approved as practicable for the tools used in the remainder of the work.

[1] Ref. D-8, Par. 335. See Bibliography, Appendix No. 4.

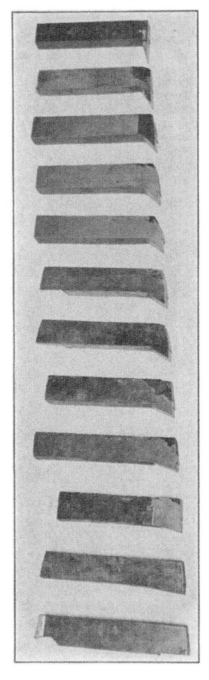

FIG. 17 TWELVE TOOLS USED IN FRONT-RAKE-ANGLE INVESTIGATION, PROBLEM (c)
From left, front-rake angles are 0, 5, 10, 15, 20, 25, 30, 35, 40, 45, 60, and 75 deg. respectively.

Front-Rake Angle — Problem (c)

37   Problem (c) is " to determine the influence of the front rake of the tool, as the variable, on the force on the tool or the energy required to remove a given volume of metal." Fig. 17 shows the 12 tools prepared for the front-rake investigation. These tools all have clearance angles of 4 deg. and a front-rake angle, starting from the left, of 0, 5, 10, 15, 20, 25, 30, 35, 40, 45, 60, and 75 deg. respectively. The data from these experiments have, for comparative purposes, also been reduced to the force in pounds per 0.001 sq. in. of cross-sectional area of the chip. The cutting speed in all cases has been confined to 20 ft. per minute, the width of cut to 0.50 in., the tool clearance to 4 deg., and the chip thicknesses to 0.006, 0.012, and 0.024 in., respectively. Representative data are shown for various metals cut. Fig. 18 shows nine of these

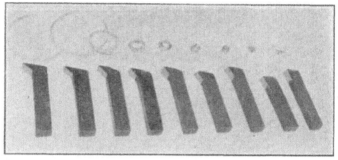

FIG. 18   NINE TOOLS HAVING FRONT RAKE ANGLES OF 45 TO 5 DEG., SHOWING STEEL CHIPS 0.500 IN. WIDE BY 0.012 IN. DEEP CUT BY EACH

tools with front-rake angles from 5 to 45 deg. and the steel chips, 0.50 in. wide by 0.012 in. deep cut by each.

38   Fig. 19 shows the unit force on the tool when cutting low-carbon steel, plotted as ordinates against the front rake angle of the tool in degrees, for three chip thicknesses. In this case the front rake was carried only to 45 deg., while the results obtained with the 0-deg. rake tool were so erratic and unreliable that they were omitted. Straight lines have been drawn through the points for each chip thickness and seem to indicate that the unit force for given conditions is reduced directly as the front-rake angle is increased. The straight line is believed to be the proper curve to represent the relation between the unit force and the front-rake angle. Were these curves extended as straight lines, they would intersect the abscissas axis within the 90-deg. front-rake-angle value. It was felt, however, that if these curves actually were continued, based on experimental results, they might extend to the right and downward but gradually become horizontal. Later

experiments confirm this. Fig. 20 gives a similar set of curves for S.A.E. 3120 steel with chip thicknesses of 0.006 and 0.012 in. These points also seem to fall reasonably well on inclined straight lines, as indicated. The curve for the 0.006-in. depth of cut would, if extended, intersect the abscissas axis to the right of the 90-deg. value. Fig. 21 shows the relation between the unit force and front-

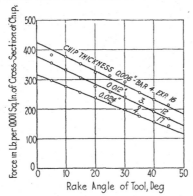

FIG. 19   FRONT-RAKE ANGLE, UNIT-FORCE CURVES FOR 0.15 PER CENT CARBON STEEL, EXPERIMENTS 16, 17, and 12

Material: 0.15 per cent carbon steel annealed.
Tools: 4 deg. clearance. Width of cut: 0.50 in.
Speed: 20 ft. per min.

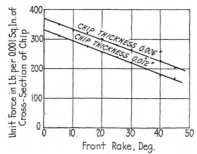

FIG. 20   UNIT FORCE — FRONT RAKE ANGLE CURVES FOR S.A.E. 3120 STEEL, EXPERIMENT 34

Material: S.A.E. 3120 steel, bar No. 31.
Tools: 4-deg. clearance. Width of cut: 0.5 in.
Speed: 20 ft. per min.

rake angle for S.A.E. 2345 steel for 0.006- and 0.012-in. thicknesses of chip. These values seem to follow the straight-line rule, but appear to decrease in value less rapidly with the increased rake angle than is true with low-carbon or S.A.E. 3120 steel. The curves for both depths of cut would, if extended as straight lines, cross the abscissas axis at some distance to the right of the 90-deg. front-rake value.

39   Fig. 22 shows several unit-force, front-rake curves for brass. Chatter occurred with heavy chips (0.018, 0.024, and 0.030 in. depth of cut) and low front-rake angles. It is interesting to note that the force was reduced for the lower rake angles as a result of the chatter. It would appear to be more economical to cut brass under conditions which cause chatter. No attempt is made

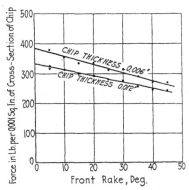

FIG. 21   UNIT FORCE — FRONT RAKE ANGLE CURVES FOR S.A.E. 2345 STEEL, EXPERIMENT 30

Material: S.A.E. 2345 steel, bar No. 29.
Tools: 4-deg. clearance. Width of cut: 0.5 in.
Speed: 20 ft. per min.

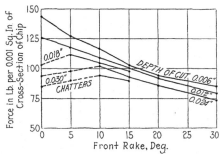

FIG. 22   UNIT FORCE — FRONT RAKE ANGLE CURVES FOR BRASS

Material: Brass, bar No. 24.
Tools: 4-deg. clearance. Width of cut: 0.506 in.
Speed: 20 ft. per min.

here to explain this, however. The points recording normal conditions for a given chip thickness seem to lie consistently on straight lines for rake angles below 20 deg. The 0.006-in. depth of cut, however, seems to have a tendency to become more horizontal between the 20- and 30-deg. front-rake angle, although with this thin chip the values obtained are less reliable than the others; the tool had a tendency to dig in on one cut and cut shallow on the next.

40   Fig. 23 shows the results of the influence of the front rake on the unit force when cutting annealed cast iron. These results seem to follow the straight-line law at least up to the 35-deg. angle and in Figs. 24 and 25 for unannealed cast iron seem to have a tendency to become more horizontal beyond the 25-deg. front rake angle. For the 0.024-in. depth of cut in Fig. 23, it appears that the curve has a tendency to become horizontal after the 35-deg. angle is reached. This is given considerable weight as the material is uniform (annealed) and for heavy cuts the chance for error is less. In Fig. 25, some results are shown with the 60-deg.

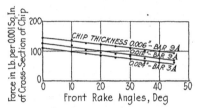

FIG. 23   UNIT FORCE — FRONT RAKE ANGLE CURVES FOR CAST IRON,
EXPERIMENT 22

Material: Cast iron, annealed.
Tools: 4-deg. clearance. Width of cut: 0.5 in.
Speed: 20 ft. per min.

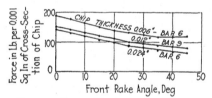

FIG. 24   UNIT FORCE — FRONT RAKE ANGLE CURVES FOR CAST IRON,
EXPERIMENT 22

Material: Cast iron, unannealed.
Tools: 4-deg. clearance. Width of cut: 0.5 in.
Speed: 20 ft. per min.

front-rake tool. It is to be noted, however, that the results for the 60-deg. front-rake angle are slightly higher than those for the 45-deg. angle. This 60-deg. tool, while cutting with some difficulty in the cast iron, cut so poorly in the case of steel that no consistent results could be obtained. The 75-deg. front-rake tool which had a lip angle of 13 deg. (equivalent to that of a razor) could not be made to cut. It jumped up and slid over the top of the metal, or dulled instantly. It was too soft even when made of high-speed steel to yield reliable results. The results shown in Fig. 25, however, add weight to the evidence that beyond a certain front-rake angle there is no further reduction in the force on the tool.

41   Fig. 26 is a summary sheet showing the influence of the front-rake angle on the unit force for all materials cut with a depth of cut of 0.012 in. It is quite obvious from these results that the relation follows the straight-line law and the force on the tool is reduced as the front-rake angle is increased, till a certain limiting angle is reached. Beyond this angle, which seems

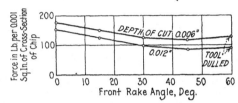

FIG. 25   UNIT FORCE — FRONT RAKE ANGLE CURVES FOR CAST IRON·
Material: Cast iron, unannealed, bar No. 1
Tools: 4-deg. clearance. Width of cut: 0.5 in.
Speed: 20-ft. per min.

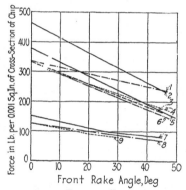

FIG. 26   SUMMARY UNIT FORCE — FRONT RAKE ANGLE CURVES
Tools: 4-deg. clearance. Depth of cut: 0.012 in. Width of cut: 0.5 in.
Speed: 20 ft. per min.
    Curve 1:   S.A.E. 2345 steel, bar No. 29.
    Curve 2:   1.03 per cent carbon steel, annealed, bar No. 32.
    Curve 3:   S.A.E. 3120 steel, bar No. 31.
    Curve 4:   S.A.E. 2320 steel, bar No. 30.
    Curve 5:   0.15 per cent carbon steel, annealed, bar No. 3.
    Curve 6:   S.A.E. 1035 steel, bar No. 28.
    Curve 7:   Cast iron, unannealed, bar No. 9.
    Curve 8:   Cast iron, annealed, bar No. 9A.
    Curve 9:   Brass, bar No. 24.

to be between 20 and 30 deg. for brass, 25 to 30 deg. for cast iron, and above 45 deg. for steel, the influence of the front rake is less pronounced. Some metals are more readily influenced by the front rake than others.

Side-Rake Angle — Problem (d)

42   Problem (d) is " to determine the influence of side rake (skew) as the variable on the force on the tool or the energy

required to remove a given volume of metal." Fig. 27 shows the
first set of tools used in a side-rake problem. They were all of
30-deg. front rake and 4-deg. clearance, but had side-rake angles
of 0, 10, 20, and 30 deg., respectively. The clearance was measured
in the direction of travel of the tool.

43 A second set of side-rake tools, each having 0-deg. front rake
and 0, 10, 20, 30, 45, 60, 75 deg. side rake, respectively, was later
used. With these tools the clearance angle was ground at right
angles to the cutting edge. The 30-deg. and 75-deg. side-rake tools,

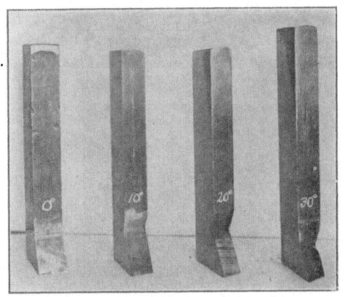

Fig. 27 First Set of Tools Used in Side-Rake Investigation,
Problem (d)

All tools have 30-deg. front rake and 0-, 10-, 20-, and 30-deg. side rake.

however, were later ground so that the clearance was measured
from the cutting edge in the direction of tool travel for check
purposes, as discussed later.

44 *First Set of Side-Rake-Angle Tools.* Fig. 28 shows a tool
having 30-deg. front rake, 30-deg. side rake, and 4-deg. clearance in the process of removing a 0.15 per cent carbon-steel
chip, which is 0.500 in. wide and 0.024 in. deep, from bar No. 6.
Fig. 29 shows a number of chips of this material which have
been cut under various conditions, the 30-deg. front-rake tools
having side-rake angles of 0, 10, 20, and 30 deg., respectively,
removing chips of 0.006, 0.012, and 0.024 in. deep and 0.500 in.

wide. All of these chips are the complete length of the bar, 36 in. It is noticeable that the diameter of the coil is increased slightly with the increase in depth of cut for a given side-rake angle. Also the diameter of the helix is reduced with the increase in side rake, but the length of the helix is increased as indicated by Table 3:

TABLE 3   MEASUREMENTS OF STEEL CHIPS PRODUCED BY SIDE-RAKE
TOOLS

Diameter and length of chip coil, in.

Side rake, deg. (30° front rake)	Depth of cut, in.		
	0.006	0.012	0.024
0	1½ dia.	1¾ dia.	2¾ dia.
10	1½ x 4	1½ x 4½	1½ x 5
20	¾ x 8	1 x 8	1¼ x 7
30	⅝ x 11	¾ x 11	1¼ x 9½

45   The slippage marks left on the land after the chip was removed were practically constant for each depth of cut for all of the tools, but decreased as the depth of cut was increased. The values as counted are shown in Table 4.

TABLE 4   SLIPPAGE ON STEEL OF SIDE-RAKE TOOLS

Depth of cut, in.	Approximate slippage
0.006	38–40 per in.
0.012	32–34 per in.
0.024	26–28 per in.

46   Fig. 30 shows a set of cast-iron chips for the various side-rake angles of the 30-deg. front-rake tools and for depths of cuts of 0.006, 0.012, and 0.024 in. The influence of the side-rake angle for each depth of cut is less marked than for the steel. Because of the brittleness of the material, the chips broke into small pieces rather than coming off in continuous lengths. For the thin chips, 0.006-in. depth of cut, the chips were longest, of the least diameter, and showed a tendency to become helical with the increase in side rake. The slippage of 0.006- and 0.012-in. depths of cuts was barely visible for any of the tools. On the 0.024-in. depth of cut, it varied from 30 slips per inch for the 0-deg. side-rake tool to 22 slips per inch for the 30-deg. side-rake tool.

47   With the first set of side-rake tools, each of which had a front-rake angle of 30 deg. as shown in Fig. 27 and side-rake angles of 0, 10, 20, and 30 deg., respectively, a series of cuts was made on brass bar No. 26. The width of cut was 0.512 in. and depths were of 0.006, 0.012, and 0.024 in., respectively. Table 5 shows the values obtained for the 0- and 30-deg. side-rake tools for all three depths of cut. The values for all four tools are given in Table 2.

48   The results show that the total force on the tool for the 0.006-in. depth of cut increased from 260 lb. for 0-deg. side rake to 300 lb. for 30-deg. side rake. This value divided by the area

of chip (0.512 multiplied by 0.006) all divided by 1000 gives the unit force or the force in pounds per 0.001 sq. in. of the cross-sectional area of the chip, which is 84.5 lb. for 0-deg. side-rake tool and 97.4 lb. for the 30-deg. side-rake tool. For all three depths

TABLE 5 TOTAL AND UNIT FORCES ON SIDE-RAKE TOOL FOR BRASS

| Depth of cut, in. | Side rake (front rake 30 deg.) | | | |
| | 0 deg. | | 30 deg. | |
	Total	Unit	Total	Unit
0.006	260	84.5	300	97.4
0.012	452	74.0	506	82.5
0.024	890	72.3	965	78.5

FIG. 28 A 30-DEG. SIDE RAKE, 30-DEG. FRONT RAKE, 4-DEG. CLEARANCE ANGLE TOOL REMOVING A 0.15 PER CENT CARBON-STEEL CHIP 0.5 IN. WIDE AND 0.024 IN. DEEP

of cut there is a very slight increase in the total and unit force on the tool. The 30-deg. front-rake tool, however, appeared to be excessive for brass as alternately uniformly thin and thick chips were cut.

49   For the 0.012-in. depth and 0.512-in. width of cut, the chips for the 0-, 10-, 20-, and 30-deg. side-rake tools were 1⅛ in. diameter spiral, ½ in. diameter by 6 in. long, $\frac{5}{16}$ in. diameter by 11½ in. long, and ¼ in. diameter and 14⅜ in. long. The original length of cut was 33 in. These values do not compare favorably with those for steel, Table 3.

50   Another series of cuts was made with the same set of tools on cast iron, bar No. 2. The width of cut was 0.500 in. and the depths of cut 0.006, 0.012, and 0.024 in. The cutting speed, as in

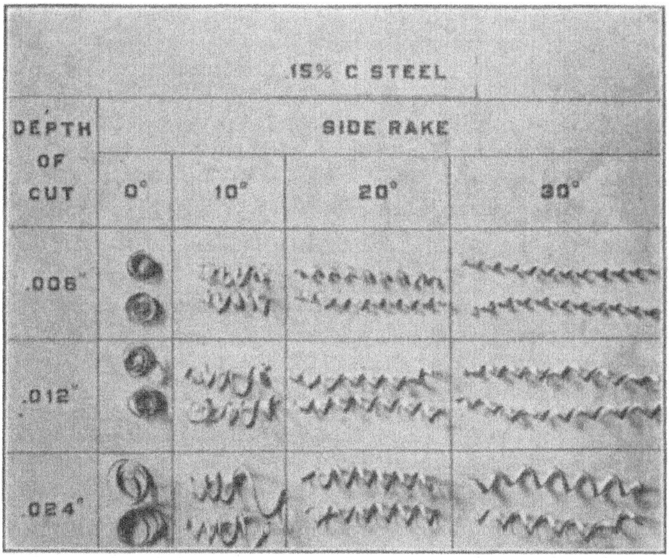

FIG. 29 0.15 PER CENT CARBON-STEEL CHIPS 0.5 IN. WIDE REMOVED BY
TOOLS HAVING 30-DEG. FRONT-RAKE AND VARIOUS SIDE-RAKE ANGLES

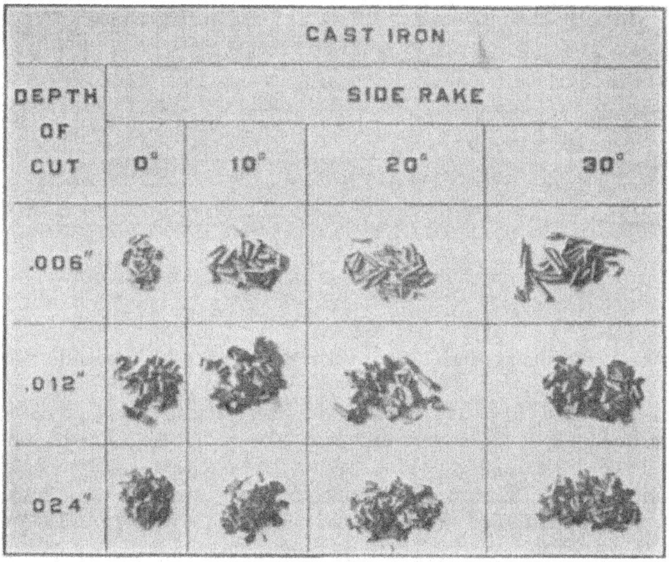

FIG. 30 CAST-IRON CHIPS 0.5 IN. WIDE REMOVED BY TOOLS HAVING
30-DEG. FRONT-RAKE AND VARIOUS SIDE-RAKE ANGLES

all other tests, was 20 ft. per min. The unit forces are shown
plotted in Fig. 31. The total force for the 0.006-in. depth of cut
is increased from 431 lb. for the 0-deg. side-rake tool to 441 lb.
for the 30-deg. side-rake tool. The corresponding values of the
unit forces as plotted are 146 and 150 lb. The increase in the unit
force for the 10-deg. side-rake tool for 0.012- and 0.024-in. depths
of cut is undoubtedly due to the fact that the tool was burned in
grinding and was considerably blunted while cutting. The cuts
were first made with a 0-deg. side-rake tool, then the 10-, 20-, and

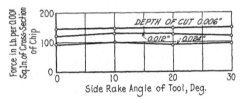

FIG. 31   UNIT FORCE — SIDE RAKE ANGLE CURVES FOR CAST IRON,
EXPERIMENT 48

Material: Cast iron, bar. No. 2.
Tools: 4-deg. clearance, 30-deg. front rake.
Width of cut: 0.5 in. Speed: 20 ft. per min.

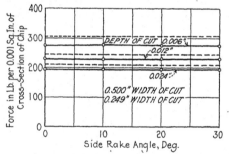

FIG. 32   UNIT FORCE — SIDE RAKE ANGLE CURVES FOR STEEL,
EXPERIMENT 49

Material: 0.15 per cent carbon steel, annealed, bar No. 6.
Tools: 4 deg. clearance, 30-deg. front rake.
Widths of cuts: 0.249 and 0.5 in. Speed: 20 ft. per min.

30-deg. tools in order. After cutting with the 30-deg. side-rake
tool, the force for the 0-deg. side-rake tool was confirmed. It
was observed that the unit force for most check tests varied not
more than 1 or 2 lb. For cast iron, the value of the check test
was about 2 lb. less, due probably to the increased softness of the
metal as the cut approached the center of the bar on account
of its being annealed in the mold.

51   A similar series of cuts was made on 0.15 per cent carbon
steel, annealed, bar No. 6. Two widths of cut, 0.249 and 0.500 in.,
respectively, were used for each depth of 0.006, 0.012, and 0.024

in. The results are shown plotted in Fig. 32. The unit forces for
the 0.249-in. width are uniformly less than those for the 0.500-in.
width. This suggests that the unit force for a side-rake-angle tool
is reduced as the width of cut is reduced and is referred to again
under the width-of-chip problem.

52   All cuts of the same depth were made on one land and run
consecutively from 0- to the 30-deg. side-rake tools and then the
0-deg. tool was again checked. On the 0.500-in. land, check tests
were made for each of the three depths of cut for the 0-deg. and
30-deg. side-rake tools with results shown in Table 6.

TABLE 6   UNIT FORCES ON SIDE-RAKE-ANGLE TOOLS FOR
0.15 PER CENT CARBON STEEL

| Depth of cut, in. | Unit force (front rake 30 deg.) | |
| | Side rake angle | |
	0 deg.	30 deg.
0.006	307	308
0.012	246	246
0.024	212	210

These values together with the intermediate values are given in
Fig. 32. With the 0.024-in. depth of cut, the tools had a tendency
to hog in, every other cut, so that readings were alternately high
and low. The values given are averages, however. These curves
appear to be straight horizontal lines.

53   Another series of tests was made on what was purchased
for S.A.E. 1035 steel, bar No. 28, but the analysis showed the
carbon content to be only 0.22 per cent. The values of the unit
forces obtained are shown in Fig. 33. The results for the 10-deg.
side-rake tool for the 0.006- and 0.012-in. depths of cut are low.
For the 0.024-in. depth of cut there appears to be a uniform
increase in the unit force from 180 lb. for the 0-deg. side-rake tool
to 194 lb. for the 30-deg. side-rake tool. This increase, however,
is very small.

54   Fig. 34 is a summary sheet showing the values of the unit
forces for the 30-deg. front-rake tools having 0-, 10-, 20-, and
30-deg. side rake for eight different materials. The width of cut
in each case was 0.500 in., the depth of cut in all cases was 0.012
in., so that all values in the figure are directly comparable. The
curve for brass has the same values as those shown in Table 6
for the 0.012-in. depth of cut. The curve for cast iron is a dupli-
cate of the middle curve in Fig. 31. The last curve is for S.A.E.
1035 steel and is a duplication of the middle curve of Fig. 33.
The curve for S.A.E. 3120 steel for the 0.012-in. depth of cut was
the only one obtained for that material. The curve for the 0.15 per
cent carbon steel is a duplication of the center dashed curve of
Fig. 32. The curve for the S.A.E. 2320 steel, bar No. 30, is the
only one obtained for this metal with this set of tools. The curve
for the S.A.E. 2345 steel, bar No. 29, is one of two curves obtained.

This curve for the 0.012-in. depth of cut shows a high value for the 20-deg. side-rake tool. Uniform results, however, were obtained with all tools for a chip depth of 0.006 in., for which the unit force for all side rake angles was 307 lb. It is, therefore, thought that the rise for the 0.012-in. depth of cut for the 20-deg. side rake angle is due to experimental error or some other outside influence. The highest curve, that for the 1.03 per cent carbon steel, is the only one obtained for this material with this set of side-rake tools.

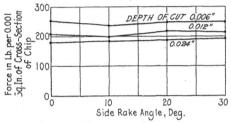

FIG. 33   UNIT FORCE — SIDE-RAKE ANGLE CURVES FOR STEEL,
EXPERIMENT 28

Material: S.A.E. 1035 steel, bar No. 28.
Tools: 4-deg. clearance, 30-deg. front rake.
Width of cut: 0.5 in. Speed: 20 ft. per min

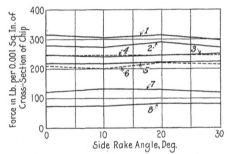

FIG. 34   SUMMARY UNIT FORCE — SIDE-RAKE ANGLE CURVES
PROBLEM (d)

Tools: 4-deg. clearance, 30-deg. front rake.
Depth of cut: 0.012 in. Width of cut: 0.5 in. Speed: 20 ft. per min.
    Curve 1: 1.03 per cent carbon steel, annealed, bar No. 32.
    Curve 2: S.A.E. 2345 steel, bar No. 29.
    Curve 3: S.A.E. 2320 steel, bar No. 30.
    Curve 4: 0.15 per cent carbon steel, annealed, bar No. 6.
    Curve 5: S.A.E. 3120 steel, bar No. 31.
    Curve 6: S.A.E. 1035 steel, bar No. 28.
    Curve 7: Cast iron, unannealed, bar No. 2
    Curve 8: Brass, rolled, bar No. 26.

55   All of the results which are available for the first set of tools, namely, those having 30-deg. front rake and 0-, 10-, 20-, and 30-deg. side rake, indicate that the value of the side-rake angle has very little influence on the unit or total force on the tool in

removing a given chip. All of the curves shown are drawn to connect the points obtained experimentally. Smooth curves through these points, however, would eliminate many of the irregularities referred to, leaving nearly all of the curves as straight horizontal lines.

56 *Second Set of Side-Rake-Angle Tools.* Because of the small cutting angles of these tools and the tendency of such angles to fail, dig in, or otherwise cause trouble, the side-rake angle of the first set of tools was not carried beyond 30 deg. The values obtained, however, would indicate that all of these curves would continue to be straight horizontal lines for side-

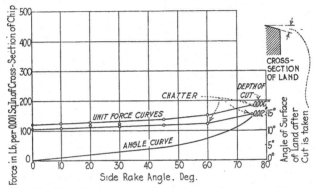

FIG. 35   UNIT FORCE — SIDE RAKE ANGLE CURVES FOR BRASS,
EXPERIMENT 57

Material: Brass, bar No. 27.
Tools: 4-deg. clearance, 0-deg. front rake.
Width of cut: 0.312 in.   Speed: 20 ft. per min.

rake angles greater than 30 deg. In order, however, to make sure of this point and to obtain comparable data with another front-rake-angle tool, the second series of side-rake tools having 0-deg. front rake was prepared.

57   Fig. 35 shows the results of the force in pounds per 0.001 sq. in. cross-sectional area of chip as a function of the side-rake angles for this set of tools when cutting brass, bar No. 27. The brass was rolled to a width of 0.312 in. Two depths of cut were taken, 0.006 and 0.012 in., respectively. The curves indicate a slight increase in the unit force for side-rake angles up to 45 deg., with a slightly accelerated increase for the 60- and 75-deg. angles. The increase up to the 45-deg. side-rake angle, however, is so small that it may be due to experimental error as smooth curves drawn through the points up to the 30-deg. angle might be horizontal. The cutting action seemed quite free for most tools. There was much chatter on the 0.012-in. depth of cut with the 60- and

75-deg. tools and some chatter on the 0.006-in. depth of cut with a 75-deg. tool.

58 The tools having the 4-deg. clearance angle ground at right angles to the cutting edge left the top of the land not horizontal but slightly inclined to the side as shown in the section, Fig. 35. A curve is plotted at the bottom showing the value of these side angles for the various side rake angles of the tool. It is less than 4½ deg. when the side-rake angle is 45 deg., but increases rapidly to 14½ deg. for a side rake angle of 75 deg. In the worst case, that for a 75-deg. side-rake-angle tool, the actual thickness of the cut is reduced to 0.9636 (cosine 14½ deg.) multiplied by the vertical depth of cut. The width of the chip would then equal the

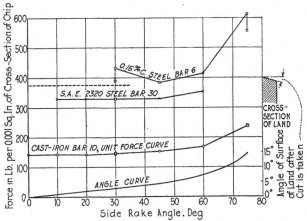

FIG. 36 UNIT FORCE — SIDE RAKE ANGLE CURVES, EXPERIMENT 58

Tools: 4-deg. clearance, 0-deg. front rake.
Width of cut: 0.5 in. Depth of cut: 0.012 in. Speed: 20 ft. per min.

horizontal width 0.312 divided by 0.9636 or 0.324 inches. This variation would appear to be of negligible influence on the force on the tool but for check purposes the 30- and 75-deg. side-rake tools were reground so that the 4-deg. clearance angle was measured in the plane of tool travel and so that the land would be horizontal after a chip was removed. The values for brass in Fig. 35 for both depths of cut were obtained with these corrected tools and found to agree with the first within a few pounds, so it was not considered necessary to regrind the other tools for this correction.

59 Fig. 36 shows the unit-force curves as a function of the side-rake angle of the 0-deg. front-rake tools for cast iron, S.A.E. 2320 steel, and 0.15 per cent carbon steel. The values for the cast iron, bar No. 10, appear to be constant at 143 lb. up to the 30-deg. side-rake angle for which 147 lb. is registered. A slight

increase to 152.5 lb. is noted for the 45-deg. angle with still greater increases to 169 and 238 lb. for the 60- and 75-deg. angles, respectively. The points marked with a circle and square for the 30- and 75-deg. side-rake tools indicate that these values were obtained with the side-rake tools before and after correcting the tools for the slope of the top of the land.

60   The curve for the S.A.E. 2320 steel, bar No. 30, appears to be horizontal until the 45-deg. side-rake angle is passed when a slight increase from 330 to 355 lb. for the 60-deg. tools is shown. The square on the curve indicates that the value obtained for the 30-deg. tool was with the tool corrected so as to leave the top of the land horizontal. This point falling on the horizontal line indicates the slight influence of the slope of the top of the land for the 30-deg. side-rake tool. The surface left on the S.A.E. 2320 steel was quite rough for the 10-deg. tool but satisfactory in all other cases. The 0-deg. front-rake and 0-deg. side-rake tool was not used because of erratic results usually obtained with this tool when cutting steel. No cuts were taken with the 75-deg. side-rake tool because of trouble experienced with the excessive side thrust on the chip which produced erratic results, and heated excessively.

61   The upper curve shows the unit forces obtained for the various side-rake angles when cutting 0.15 per cent carbon steel, bar No. 6. The value of the unit force for side-rake angles below 30 deg. is shown by a straight dashed line. Actual test values for the 0-deg. front and side-rake tool were so erratic that they were thought unreliable. The value of 375 lb. as the unit force was taken from the front-rake investigation curves, Fig. 19. The values represented by the solid line were obtained with difficulty and are not considered wholly reliable. The points for the 30-deg. and 75-deg. side-rake-angle tools were checked with the tools corrected for clearance, the new values being lower as indicated by the square. It is interesting to note that in Fig. 36 the unit force for the 0.15 per cent carbon steel is 375 lb. and for S.A.E. 2320 steel only 330 lb. In Fig. 34, where the 30-deg. front-rake tools were used, the values for the unit forces of the two materials are almost equal at 245 lb., the same bars being used in each case.

62   The values presented for both sets of side-rake tools for the various metals cut seem to indicate decisively that for moderate values, the side-rake angle, all other factors remaining constant, has no influence on the total or unit force on the tool. When the side-rake angle reaches 60 deg., a definite increase is noted and for greater values of side rake the force is markedly increased.

Force — Chip Width, and Depth of Cut, Problem (e)

63   Problem (e) is " to determine the influence of the depth of cut or width of cut, as the variable, on the force on the tool or the energy required to remove a given volume of metal." In order

to have but one variable at a time, the problem has been divided into two parts so as to determine, first, the influence on the force on the tool of a change in width of chip, and, secondly, the influence on the force due to a change in the depth of cut.

64   *Force — Chip Width.* The 30-deg. front-rake tool having a 4-deg. clearance angle was found to give consistently reliable results in previous tests on most materials. It was therefore selected as the tool to be used in the chip-width series of tests.

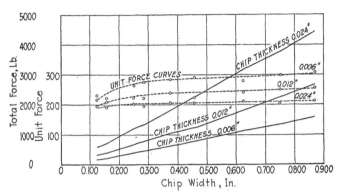

FIG. 37   UNIT AND TOTAL FORCE — CHIP WIDTH CURVES FOR STEEL, EXPERIMENT 20

Material: 0.15 per cent carbon steel, annealed, bar No. 5.
Tool: 4-deg. clearance, 30-deg. front rake. Speed: 20 ft. per min.

For brass, however, in order to avoid chatter, which occurs in taking heavy cuts with small-front-rake-angle tools, and to avoid hogging in, or taking of alternately thick and thin chips, which occurs with tools of excessive front rake, such as the 30-deg. front-rake tool, a tool having 15-deg. front rake and 4-deg. clearance was used. A 0-deg. front-rake tool was used with the 30-deg. tool on annealed cast iron and on brass as noted.

65   The usual cutting speed was 20 ft. per min. This is rather low for cutting brass. A number of check runs were made at 35 ft. per min. for brass as well as some of the steels, and identical results were obtained in every case for the two speeds.

66   Depths of cut were confined in most instances to 0.006, 0.012, and 0.024 in. The width of cut ranged between ⅛ in. and 1 in. Fig. 37 shows the total- and unit-force curves for various chip widths for 0.15 per cent carbon steel, annealed, bar No. 5. The total-force curves are shown as solid lines, the lines having been drawn between the actual experimental points rather than as a smooth curve. The total-force curves appear to be straight

lines, although they are slightly concave upward. The slight variations are due probably to experimental error. If the curves were extended to a chip having zero width, they would pass through the origin. This point is discussed later. The unit-force curves shown dashed are drawn as smooth curves through the points as the points are well scattered. A slight variation of the total force from the smooth curve shows up considerably magnified in the unit-force curve which is plotted on a much larger scale. It appears that the unit-force curves to the right of 0.3-in. chip width are nearly horizontal, but to the left they have a tendency to drop. It was difficult to secure constant readings with a narrow chip width, as a small error in the reading of the total force represented a large percentage of the reading. However, the values for the

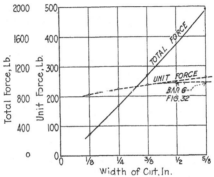

FIG. 38   UNIT AND TOTAL FORCE — CHIP WIDTH CURVES

Material: 0.15 per cent carbon steel, bar No. 5.
Tool: 4-deg. clearance, 30-deg. front rake, 20-deg. side rake.
Depth of chip: 0.012 in.

0.158-in. width are believed to be a little too low. The values for a chip width of ⅛ in. are high for all three depths of cut.

67   In Fig. 32 it was noticed that the unit forces for side-rake tools on a land of 0.5 in. were higher than those for the 0.25-in. width. As a check, a tool having 30-deg. front-rake and 20-deg. side-rake angle was selected for a test on a 0.15 per cent carbon steel, bar No. 5, the depth of cut being 0.012 in. and the width ⅛, ¼, ⅜, and ⅝ in. respectively, as these lands were already prepared. The results of this test are shown in Fig. 38, the solid line representing the total-force curve and the dashed line, the unit-force curve. The total-force curve appears to be slightly concave upward, causing the unit-force curve to be slightly concave downward for the ⅛- and ¼-in. widths. The value of the total forces for the ⅜-in. width of cut may be slightly high, as it was difficult to obtain consistent readings. Two points were taken from Fig. 32, bar No. 6, for the ½- and ⅝-in. widths, which also agree with the values in Fig. 37, and the unit-force curve drawn through them.

The unit-force curve is unquestionably lower with the narrow widths of cut and has a tendency to become horizontal with the greater widths of cut. The 235 lb. for the ¼-in. width and the 245 lb. for the ½-in. width check very well with the unit forces for the 0.012-in. depth of cut on Fig. 32. The finish left on the lands after a chip had been removed seemed to be finer for the narrower widths of cut. Slippage lines were counted as follows: 40 to 60 for the ⅛-in. width; 33 to 40 for the ¼-in. width; 28 to 32 for the ⅜-in. width; 24 to 30 for the ⅝-in. width; and about 12 to 20 for the ¾-in. width, the force values for which were too erratic to plot. The first number of these slippage values corresponds to the continuous lines across the land at an angle of about 20 deg. The second and larger figure corresponds to the number of lines countable, some of which were not continuous.

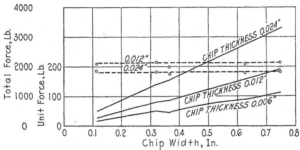

FIG. 39   UNIT AND TOTAL FORCE — CHIP WIDTH CURVES,
EXPERIMENT 29

Material: S.A.E. 1035 steel, bar No. 28.
Tool: 4-deg. clearance, 30-deg. front rake. Speed: 20 ft. per min.

The chip coils in each case were ¾ in. outside diameter by 2½ in. long for the ⅛-in. land, 1¾ by 3½ in. for the ¼-in. land, 2 by 3¾ in. for the ⅜-in. land, 3 by 4½ in. for ⅝-in. land (chip was conical) and 3 by 5 in. for the ¾-in. land (chip was conical).

68   Fig. 39 shows the total and unit forces for various chip widths for S.A.E. 1035 steel, bar No. 28. The total-force curve for the 0.006-in. chip is somewhat irregular, probably due to experimental error, associated with light cuts, therefore, unit-force curves have been drawn for only the 0.012- and 0.024-in. depths of cut. Both unit-force curves appear to be horizontal straight lines with no tendency to dip for the narrow chip widths. Fig. 40 shows the total and unit forces obtained for the various chip widths for 1.03 per cent carbon steel, annealed, bar No. 32. The total-force curve is obviously a straight line increasing directly with the width of cut. It is noted here that the curve when extended to the axis of abscissas as a straight line, passes not through the origin, but crosses the axis a short distance to the

right. This accounts for the tendency of the unit-force curve, shown in dotted lines, to dip with the low values of chip width. Several of the following figures for steel show the total-force curve to be similar to this one. All steel chips were noticed to have on the side which slides over the tool face a narrow band at either edge which appeared somewhat smoother than the surface in between. It was thought the intensity of stress, due to the removal of the chip by the tool, was less on these two narrow bands than

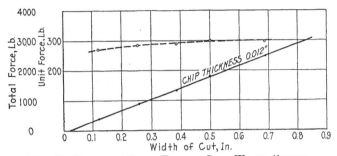

FIG. 40   UNIT AND TOTAL FORCE — CHIP WIDTH CURVES,
EXPERIMENT 45

Material: 1.03 per cent carbon steel, annealed, bar No. 32.
Tool: 4-deg. clearance, 30-deg. front rake. Speed: 20 ft. per min.

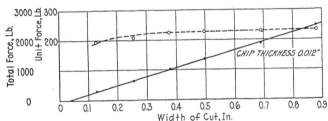

FIG. 41   UNIT AND TOTAL FORCE — CHIP WIDTH CURVES,
EXPERIMENT 41

Material: S.A.E. 2320 steel, bar No. 30.
Tool: 4-deg. clearance, 30-deg. front rake. Speed: 20 ft. per min.

on the portion of the chip between them. As the width of cut approaches zero the total force would approach zero so that all of these total-force curves should pass through the origin by turning slightly concave upwards for the very thin chips. The reason for these edges on the steel chips cannot be accounted for unless it is due to lubrication. In milling the steel test bars in a milling machine, they were lubricated with lard oil and although wiped dry and sometimes cleaned with gasoline, there may have been a trace of oil left. This trace, so small as not to be seen, would cause some smoke during a heavy cut. It may be that the small amount of oil actually lubricated the chip at either side.

69 Fig. 41 shows a similar total- and unit-force curve for S.A.E. 2320 steel, bar No. 30. The unit force appears to be practically horizontal until a width of cut of less than ¼ in. is reached. Fig. 42 shows a total- and unit-force curve for S.A.E. 2345, bar No. 29. The 30-deg. front-rake tool is used in all cases. It is only for a chip width of ⅛ in. that the unit force drops perceptibly, although it is probably gradually reduced from the ¼-in. width. Fig. 43 shows similar curves for the S.A.E. 3120 steel, bar No. 31. The unit-force curve here slightly slopes downward to the left.

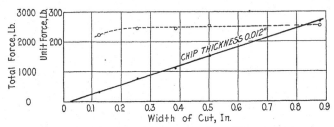

FIG. 42 UNIT AND TOTAL FORCE — CHIP WIDTH CURVES, EXPERIMENT 33

Material: S.A.E. 2345 steel, bar No. 29.
Tool: 4-deg. clearance, 30-deg. front rake. Speed: 20 ft. per min.

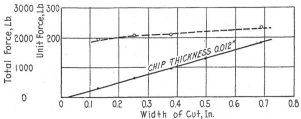

FIG. 43 UNIT AND TOTAL FORCE — CHIP WIDTH CURVES, EXPERIMENT 37

Material: S.A.E. 3120 steel, bar No. 31.
Tool: 4-deg. clearance, 30-deg. front rake. Speed: 20 ft. per min.

The value for the ⅛-in. width, however, is not far below the values for greater widths.

70 Values for three depths of cut for cast iron, annealed, bar No. 7A, are shown in Fig. 44. A 0-deg. front-rake tool was used. The narrowest chip cut was ¼ in. in width. The values of the unit curves appear to be horizontal except for the points at the extreme right for a width of cut of 0.775 in. where a distinct increase is noted. This is attributed to some experimental error, as the increase occurs for all three depths of cut for that width only. Fig. 45 shows a similar set of curves for the same annealed cast iron, bar No. 7A, for a 30-deg. front-rake tool. While the unit

values are somewhat lower than those for similar depths of cut in Fig. 44, they have the same characteristics except for the increase for the 0.775-in. width of cut. The values for the ¼-in. width of cut in Fig. 45 for the 0.012- and 0.024-in. depths of cut appear slightly above the normal unit-force curve. This is probably due to experimental error or variation in material, which for cast iron is a common source of annoyance.

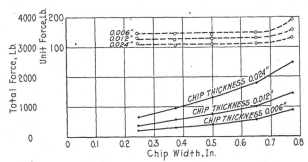

FIG. 44   UNIT AND TOTAL FORCE — CHIP WIDTH CURVES,
EXPERIMENT 25

Material: Cast iron, annealed, bar No. 7A.
Tool: 4-deg. clearance, 0-deg. front rake. Speed: 20 ft. per min.

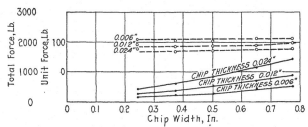

FIG. 45   UNIT AND TOTAL FORCE — CHIP WIDTH CURVES,
EXPERIMENT 25

Material: Cast iron, annealed bar No. 7A.
Tool: 4-deg. clearance, 30-deg. front rake. Speed: 20 ft. per min.

71   Fig. 46 shows three total- and unit-force curves for cast iron, unannealed, bar No. 11. The tool used in this case had 0-deg. front rake. The results given in Fig. 47 are for the same material, but with a tool having 30-deg. front rake. The values are greater for the 0-deg. front-rake tool as expected. The characteristics of the unit-force curves, however, still indicate no influence due to the change in width of chip when the depth remains constant. The unit-force curves for the 0.024-in. depth of cut in both figures would turn downward to the left as the total-force curve must be concave upwards in order to pass through the origin.

72 Figs. 48 and 49 show total and unit forces for brass, bars
Nos. 33 and 34, respectively, the tool in each case ha.ing 15-deg.
front rake. Values for the unit-force curves for the 0.006- and
0.024-in. depth of cut only are shown in both cases, as the unit-
force values for intermediate depths of cut would lie in between
and cause congestion. The total-force curves in both figures
appear to be straight lines passing through the origin and, therefore, the unit-force curves would continue horizontally to the left.

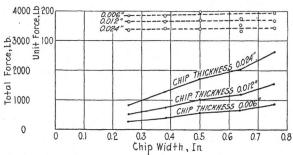

FIG. 46 UNIT AND TOTAL FORCE — CHIP WIDTH CURVES,
EXPERIMENT 25

Material: Cast iron, unannealed, bar No. 11.
Tool: 4-deg. clearance, 0-deg. front rake. Speed: 20 ft. per min.

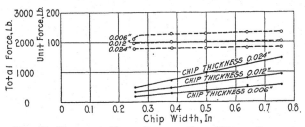

FIG. 47 UNIT AND TOTAL FORCE — CHIP WIDTH CURVES,
EXPERIMENT 25

Material: Cast iron, unannealed, bar No. 11.
Tool: 4-deg. clearance, 30-deg. front rake. Speed: 20 ft. per min.

73 Fig. 50 is a summary sheet of total forces for a depth of
cut of 0.012 in. for various widths of cut for all of the materials
used. A 30-deg. front-rake tool was used in all cases except for
curve 7 for brass, for which a 15-deg. front-rake tool was used.
The total-force curves, which are concave upward, cause the unit-
force curves to be concave downward and to drop for small widths
of cut, also, those total-force curves which cross the abscissa
axis to the right of the origin, cause the unit-force curve to drop
for small widths of cut. Fig. 51 is another summary sheet showing
the unit forces for the curves of Fig 50, plotted over the widths

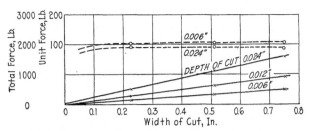

FIG. 48   UNIT AND TOTAL FORCE — CHIP WIDTH CURVES,
EXPERIMENT 56

Material, Brass, bar No. 33.
Tool: 4-deg. clearance, 15-deg. front rake. Speed: 20 ft. per min.

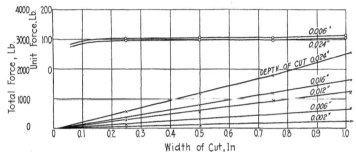

FIG. 49   UNIT AND TOTAL FORCE — CHIP WIDTH CURVES,
EXPERIMENT 56

Material: Brass, bar No. 34.
Tool: 4-deg. clearance, 15-deg. front rake. Speed: 20 ft. per min.

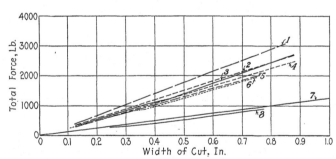

FIG. 50   SUMMARY OF TOTAL FORCE — CHIP WIDTH CURVES,
PROBLEM (e)

Tools: 4-deg. clearance. Depth of cut: 0.012 in. Speed: 20 ft. per min.
Curve 1: 1.08 per cent carbon steel, bar No. 32, front-rake angle 30 deg.
Curve 2: S.A.E. 2345 steel, bar No. 29, front-rake angle 30 deg.
Curve 3: 0.15 per cent carbon steel, bar No. 5, front-rake angle 30 deg.
Curve 4: S.A.E. 2320 steel, bar No. 30, front-rake angle 30 deg.
Curve 5: S.A.E. 3120 steel, bar No. 31, front-rake angle 30 deg.
Curve 6: S.A.E. 1035 steel, bar No. 28, front-rake angle 30 deg.
Curve 7: Brass, bar No. 34, front-rake angle 15 deg.
Curve 8: Cast iron, annealed, bar No. 7A, front-rake angle 30 deg.

of cut as abscissas. All curves except 6 for S.A.E. 1035 steel and 9 for annealed cast iron are shown concave downward for the small widths of cut. Both curves would, however, turn downward if continued to the left as the total-force curves in Fig. 39 and 45 would have to turn concave upward in order to pass through the origin. From this evidence it is concluded that the total and unit forces are lower for the narrower width of cut. The unit force increases rapidly with an increase in width of cut for most metals until a width of 0.24 in. is reached, after which the increase is less and in some instances there is no increase.

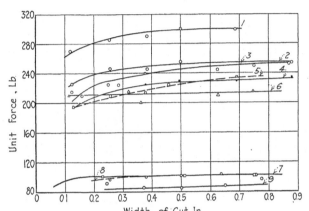

Fig 51 Summary of Unit Force — Chip Width Curves, Problem (e)

Tools: 4-deg. clearance. Front-rake angle 30 deg. (except brass, 15 deg.).
Depth of cut: 0.012 in. Speed: 20 ft. per min.
Curve 1: 1.03 per cent carbon steel, bar No. 32.
Curve 2: S.A.E. 2345 steel, bar No. 29.
Curve 3: 0.15 per cent carbon steel, bar No. 5.
Curve 4: S.A.E. 2320 steel, bar No. 30.
Curve 5: S.A.E. 3120 steel, bar No. 31.
Curve 6: S.A.E. 1035 steel, bar No. 28.
Curve 7: Brass, rolled, bar No. 33.
Curve 8: Cast iron, unannealed, bar No. 11.
Curve 9: Cast iron, annealed, bar No. 7A.

74 Force — Depth of Cut. The results of the second part of Problem (e), i.e., to show the influence on the force on the tool due to a change in the depth of cut or chip thickness for a constant width of cut, are given below. Again the 30-deg. front-rake tool having 4-deg. clearance was used for all metals except brass, when the 0-deg. and 15-deg. front-rake tools were used. A 0-deg. tool was used with the 30-deg. front-rake tool on cast iron as is noted.

75 Fig. 52 shows the total- and unit-force curves for 0.15 per cent carbon steel, bar No. 4. The width of cut is $\frac{1}{8}$ in. and the depth of cut varies between 0 and 0.100 in. The total-force curve has been drawn between the actual experimental points and ap-

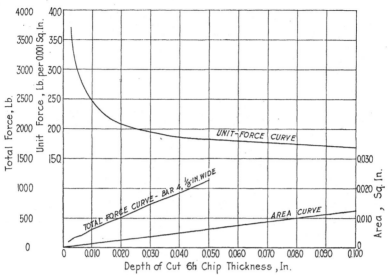

FIG. 52   CHIP-THICKNESS CURVES, EXPERIMENT 14

Material: 0.15 per cent carbon steel, bar No. 4.
Tool: 4-deg. clearance, 30-deg. front rake.
Chip width: ⅛ in. Speed: 20 ft. per min.

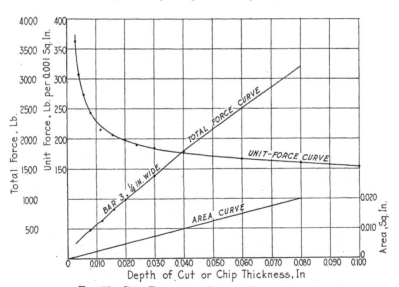

FIG. 53   CHIP-THICKNESS CURVES, EXPERIMENT 18

Material: 0.15 per cent carbon steel, bar No. 3.
Tool: 4-deg. clearance, 30-deg. front rake.
Chip width: ¼ in. Speed: 20 ft. per min.

pears to be somewhat irregular. If continued to the left and downward, it would pass through the coördinate axis, but would have to dip down considerably. The curve is obviously concave downward. Values for extremely small depths of cut were obtained with difficulty and in many cases proved unreliable. An error of a small fraction of a thousandth of an inch in depth of cut for such small depths of cut proved to be excessive. A slight error in reading the gage would also cause a wide variation in the total and unit force. The unit-force curve is shown plotted in the upper part of Fig. 52 and appears to be nearly horizontal for great

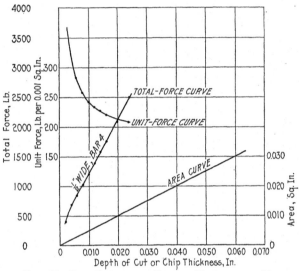

FIG. 54   CHIP-THICKNESS CURVES, EXPERIMENT 13

Material: 0.15 per cent carbon steel, bar No. 4.
Tool: 4-deg. clearance, 30-deg. front rake.
Width of cut: 0.500 in. Speed: 20 ft. per min.

depths of cut, but turns upward as the depth of cut is reduced below 0.030 in. Fig. 53 shows a similar total- and unit-force curve for 0.15 per cent carbon steel, bar No. 3, the width of land being ¼ in. The corresponding values for a ½-in. width of land on bar No. 3 are shown in Fig. 54, while the values in Fig. 55 are produced from a 1-in. width of land on bar No. 4. ·

76   For the sake of comparison, the total-force curves, on the 0.15 per cent low-carbon steel for all four chip widths are shown in Fig. 56. The curves for the ⅛-, ½-, and 1-in. widths being of bar No. 4, are comparable. The total force for a 0.012-in. depth of cut for each of the three widths is 350, 1400, and 2725 lb., respectively. The value for the ¼-in. width of bar No. 3 is 655 lb. and is slightly lower than a smooth curve through the three values

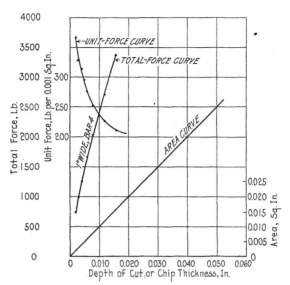

FIG. 55   CHIP-THICKNESS CURVES, EXPERIMENT 13

Material: 0.15 per cent carbon steel, bar No. 4.
Tool: 4-deg. clearance, 80-deg. front rake.
Width of cut: 1 in.   Speed: 20 ft. per min.

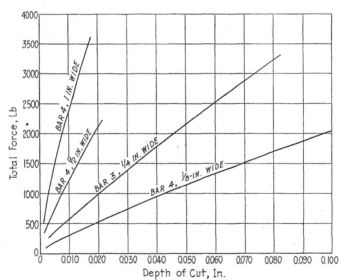

FIG. 56   CHIP-THICKNESS CURVES

Material: 0.15 per cent carbon steel.
Tool: 4-deg. clearance, 80-deg. front rake.   Speed: 20 ft. per min.

for bar No. 4 plotted over their chip widths. Fig. 57 shows the unit-force curves for the total-force curves of Fig. 56 for the four widths of cut of the 0.15 per cent carbon steel. The depth of cut was 0.012 in. and the 30-deg. front-rake tool was used. These unit curves indicate the tendency of the unit forces to increase as the depth of cut is reduced for all widths of cut. The ordinate scale is large and the curves, particularly for bar No. 4, are close together for ⅛-, ½-, and 1-in. widths.

77 Fig. 58 shows a total- and unit-force curve as a function of the depth of cut for S.A.E. 1035 steel, bar No. 28. The value for

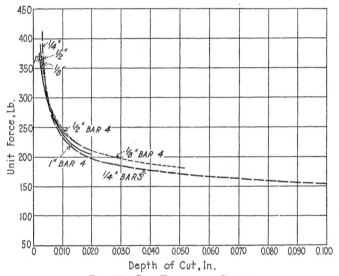

FIG. 57 CHIP-THICKNESS CURVES

Material: 0.15 per cent carbon steel, bars Nos. 3 and 4.
Tool: 4-deg. clearance, 30-deg. front rake. Speed: 20 ft. per min.

both the total and unit forces for the 0.003 in. depth of cut is obviously in error. Otherwise, the curves are similar in character to those for 0.15 per cent carbon steel. Fig. 59 shows a similar set of curves for the 1.03 per cent carbon steel, bar No. 32; Fig. 60 for S.A.E. 2320 steel; bar No. 30; Fig. 61 for S.A.E. 2345 steel, bar No. 29; and Fig. 62 for S.A.E. 3120 steel, bar No. 31. In all of the curves for the steel, there is a distinct tendency for the unit force to curve upward for depths of cut below 0.024 in. For depths greater than this, the curves become more and more horizontal, but seem to slope to the right for all depths.

78 Values for the total and unit forces when cutting cast iron annealed, bar No. 8A with both the 30- and 0-deg. front-rake tools, are shown in Fig. 63. The total-force curve for the 0-deg. front-

rake tool is considerably higher than that for the 30-deg. front-rake tool. The curves, however, appear to be slightly concave downward, becoming more so with the reduced depth of cut. The unit forces for the 0-deg. front-rake tool are also higher than

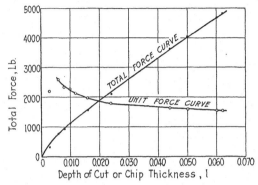

FIG. 58   CHIP-THICKNESS CURVES, EXPERIMENT 27

Material: S.A.E. 1035 steel, bar No. 28.
Tool: 4-deg. clearance, 30-deg. front rake. Width of cut: 0.5 in.
Speed: 20 ft. per min.

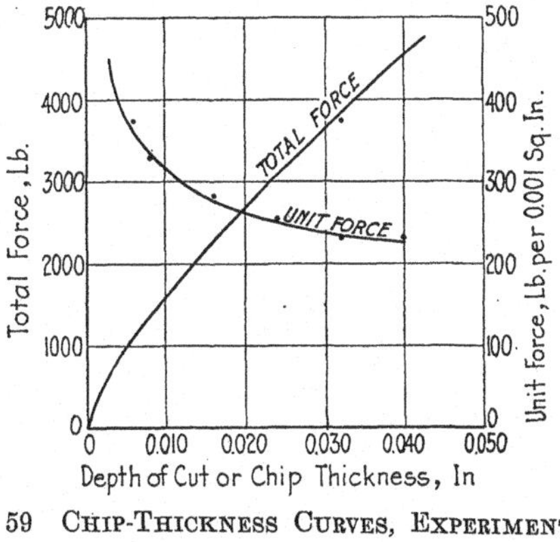

FIG. 59   CHIP-THICKNESS CURVES, EXPERIMENT 44

Material: 1.03 per cent carbon steel, bar No. 32.
Tool: 4-deg. clearance, 30-deg. front rake.
Width of cut: 0.5 in. Speed: 20 ft. per min.

those for the 30-deg. front-rake tool. However, both turn upward decidedly for depths of cut less than 0.024 in. Similar values are shown for unannealed cast iron, bar No. 8, in Fig. 64. These values appear more uniform than those for the annealed bar. The total-force curves for both the 0- and 30-deg. front-rake-angle tools have a slightly greater curvature concave downward than for the annealed bar. The characteristics are quite similar. The unit-force curves, as for the annealed cast iron are lowest for the greatest depth of cut, are concave upward at all points, and are accelerated upward with a reduction in depth of cut. Below 0.025-in. depth of cut, there is a marked change in the unit force for a given change in depth of cut, while for values above 0.025 in., the change is less marked.

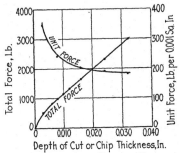

FIG. 60   CHIP-THICKNESS CURVES, EXPERIMENT 40

Material: S.A.E. 2320 steel, bar No. 30.
Tool: 4-deg. clearance, 30-deg. front rake.
Width of cut: 0.5 in.   Speed: 20 ft. per min.

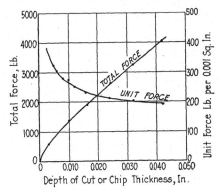

FIG. 61   CHIP-THICKNESS CURVES, EXPERIMENT 32

Material: S.A.E. 2345 steel, bar No. 29.
Tool: 4-deg. clearance, 30-deg. front rake.
Width of cut: 0.5 in.   Speed: 20 ft. per min.

79   Fig. 65 shows the total forces on 15-deg. front-rake tool when cutting brass, bar No. 34, for widths of cut of $\frac{1}{4}$, $\frac{1}{2}$, $\frac{3}{4}$, and 1 in., respectively. All curves are drawn as straight lines and seem to fit the conditions quite well. The curves if extended as straight lines would cross the ordinate axis above the zero point. They would, however, as true curves, turn downward so as to pass through the origin when the depth of cut is zero. The values for the total force for the 0.024-in. depth of cut are 570, 1175, 1800,

and 2570 lb., respectively. When plotted over the widths of chip as abscissas, a curve slightly concave upward is obtained as shown by the total-force curve in Fig. 49. It is for this reason that the unit-force curves in Figs. 48 and 49 turn down for small widths of cut. Two unit-force curves are drawn in Fig. 65 for the 1.015-in. and 0.250-in. widths. No intermediate values are shown, as they lie in between these two curves which in themselves are close together. The unit-force curves show the usual tendency to turn upward for small values of depth of cut.

80   Fig. 66 shows the total-force curves for 0.006-, 0.012-, and 0.024-in. depths of cut, for four widths of cut. Each width, however, is from a different bar, so that the results are not directly comparable. The unit-force values for the 1.015-in. width, bar

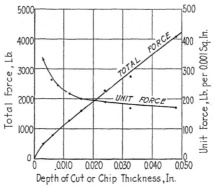

Fig. 62   Chip-Thickness Curves, Experiment 36

Material: S.A.E. 3120 steel, bar No. 31.
Tool: 4-deg. clearance, 30-deg. front rake.
Width of cut: 0.5 in.   Speed: 20 ft. per min.

No. 34, are 134 lb. for the 0.006-in. chip and 123 lb. for the 0.012-in. depth of cut. The unit-force values for the other curves also show those for the 0.006-in. depth of cut to be greater than those for the 0.012-in. depth of cut, indicating that the unit-force curve would turn upward for the thinner chips. The curves shown are drawn through the actual experimental points. Notation is made for those points at which chatter occurred. When chatter occurs the force on the tool is reduced.

81   Fig. 67 shows the total-force curves for a 0.308-in. width of cut, bar No. 27, and a 0.512-in. width of cut, bar No. 26, plotted on depth of cut in inches. The lower portions of the curves have been drawn as straight lines and would cross the ordinate axis, if continued as straight lines, slightly above the origin. The dashed lines indicate the total-force curves if continued upward as straight lines without chatter. The full lines are drawn through the actual experimental points. The points at which chatter occurred are

indicated. Unit-force curves are shown for both widths of cut
for the straight-line, total-force curves. That for the 0.512-in.
width is shown above that for the 0.308-in. width. While it is
expected that the unit-force curve for the wider chip will be
higher, the marked difference in values is accounted for in this
case by the difference in materials.

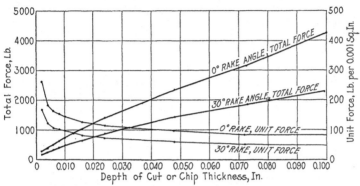

FIG. 63 CHIP-THICKNESS CURVES, EXPERIMENT 23

Material: Cast iron, annealed, bar No. 8A.
Tools: 4-deg. clearance. Width of cut: 0.5 in. Speed: 20 ft. per min.

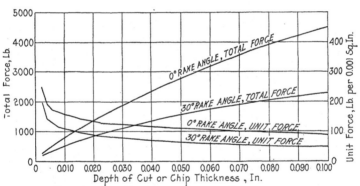

FIG. 64 CHIP-THICKNESS CURVES, EXPERIMENT 23

Material: Cast iron, unannealed, bar No. 8.
Tools: 4-deg. clearance. Width of cut: 0.5 in. Speed: 20 ft. per min.

82 Fig. 68 shows all total-force curves for the 0.5-in. width
of cut for the nine different materials cut, plotted on the depths
of cut as abscissas: All of the curves are decidedly concave downward with increased curvature for the thinner chips. Curve 7
for brass is drawn as a straight line, although it is actually very
slightly concave downward. A 30-deg. front-rake tool was used
in all cases except for brass, for which a 15-deg. front-rake tool

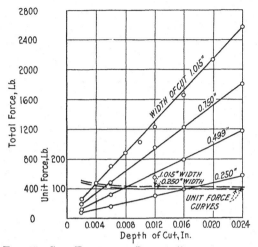

FIG. 65 CHIP-THICKNESS CURVES, EXPERIMENT 56

Material: Brass, bar No. 34.
Tool: 4-deg. clearance, 15-deg. front rake. Speed: 20 ft. per min.

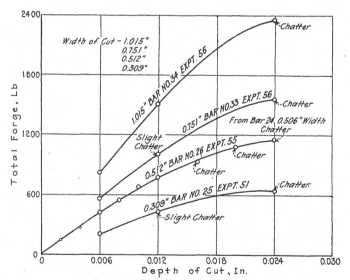

FIG. 66 CHIP-THICKNESS CURVES

Material: Brass, bars Nos. 25, 26, 33, and 34.
Tool: 4-deg. clearance, 0-deg. front rake. Speed: 20 ft. per min.

was used. Corresponding values for the unit forces for the various metals cut, plotted on the depths of cut in inches as abscissas, are shown in Fig. 69. These curves show that the unit forces for the very thin chips (0.003 in. and less) are relatively high, but are reduced rapidly with an increase in depth of cut. This reduction in unit force becomes more retarded as the depth of cut is increased so that for depths of cut of 0.030 in. or greater, there appears to be little reduction in unit force for an increase in depth of cut.

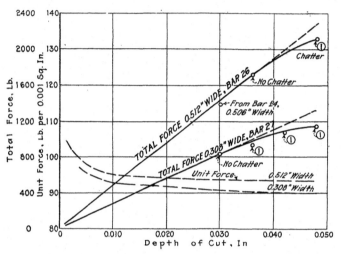

FIG. 67   CHIP-THICKNESS CURVES, EXPERIMENT 52

Material: Brass, bars Nos. 26 and 27.
Tool: 4-deg. clearance, 15-deg. front rake. Speed: 20 ft. per min.

Physical Properties of Materials Cut — Problem (f)

83   Problem (f) is " to find the relation between the force on the tool of a given shape required to remove a specific chip of a given material and the physical or chemical properties of the material." For this purpose, the unit force, that is, the force per 0.001 sq. in. of cross-sectional area of chip for a 30-deg. front-rake tool, has been selected for each material to compare with the other physical properties of the material. The unit force for the 30-deg. tool was selected, as that tool gave satisfactory results for all metals except brass. Even that for brass with the 0.012-in. depth of cut is considered reliable and representative of the properties of the material. All unit forces were determined from a cut 0.5 in. wide and 0.012 in. deep. These unit forces are shown in Fig. 26.

84   In Fig. 70 several curves are plotted to show any relation which might exist between the unit forces and the other physical properties of the materials. The abscissas 1, 2, 3, etc., represent

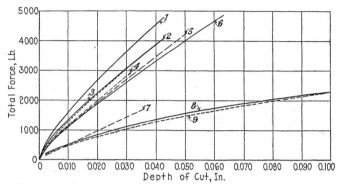

FIG. 68   SUMMARY OF TOTAL FORCE — CHIP THICKNESS CURVES
PROBLEM (e)

Tools: 4-deg. clearance, 30-deg. front rake (except brass, 15-deg.)
Width of cut: 0.5 in. Speed: 20 ft. per min.
Curve 1: 1.03 per cent carbon steel, bar No. 32.
Curve 2: S.A.E. 2345 steel, bar No. 29.
Curve 3: 0.15 per cent carbon steel, bar No. 4.
Curve 4: S.A.E. 2320 steel, bar No. 30.
Curve 5: S.A.E. 3120 steel, bar No. 31.
Curve 6: S.A.E. 1035 steel, bar No. 28.
Curve 7: Brass, rolled, bar No. 26.
Curve 8: Cast iron, unannealed, bar No. 8.
Curve 9: Cast iron, annealed, bar No. 8A.

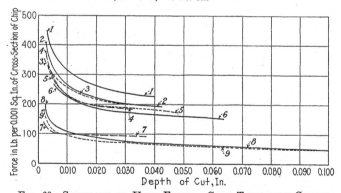

FIG. 69   SUMMARY OF UNIT FORCE — CHIP THICKNESS CURVES

Tools: 4-deg. clearance, 30-deg. front rake (except brass, 15-deg.)
Width of cut: 0.5 in. Speed: 20 ft. per min.
Material: Curve numbers same as in Fig. 68.

the various materials referred to in the table in the figure. These materials are representative of all of those cut and are arranged in order of intensity of unit forces. Curve A joins the unit-force values for each material. The values decrease with each material from left to right. Curves *B*, *C*, and *D* indicate the Brinell, Rockwell, and scleroscope hardness numbers, respectively, for each material. The Brinell numbers for the steels 1, 3, and 6 are low on

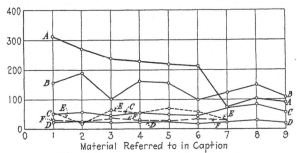

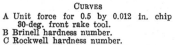

FIG. 70   PHYSICAL-PROPERTY CURVES

CURVES	MATERIALS
A Unit force for 0.5 by 0.012 in. chip 30-deg. front rake tool.	1 1.03 per cent carbon steel, bar No. 32.
B Brinell hardness number.	2 S.A.E. 2345 steel, bar No. 29.
C Rockwell hardness number.	3 0.15 per cent carbon steel, bar No. 4.
D Scleroscope hardness number.	4 S.A.E. 2320 steel, bar No. 30.
E Percentage reduction of area (tension).	5 S.A.E. 3120 steel, bar No. 31
F Percentage elongation in 2 inches.	6 S.A.E. 1035 steel, bar No. 28.
	7 Brass, rolled, bar No. 26.
	8 Cast iron, unannealed, bar No. 8.
	9 Cast iron, annealed, bar No. 8A.

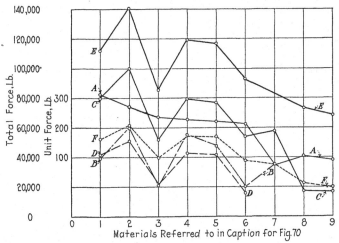

FIG. 71   PHYSICAL-PROPERTY CURVES

(See Par. 85. Material number same as in Fig. 70.)

the curve, while the corresponding unit forces remain relatively even with the others. The unit forces for brass, 7, unannealed cast iron, 8, and annealed cast iron, 9, are less than half of those for steel and are even greater than those for steels 3 and 6. It is appreciated that for other heat treatments of the materials the unit force and hardness numbers would probably be different. There appears to be no helpful relation between the unit-force and the hardness-number curves. The highest point on the Rockwell and scleroscope hardness curves is for cast iron, 8. Curves E and F shown in dashed lines represent, in per cent, the reduction of area and elongation in 2 in., respectively, for each material. The reduction of area appears to be highest for the steels which have the lowest unit forces except the S.A.E. 2345 steel, 2, but drops with the unit force for brass, 7, as the unit force also does. No relation between the unit force and these factors seems to be suggested.

85   Fig. 71 shows the unit force and strength plotted over the materials referred to in the table in Fig. 70 as abscissas. The

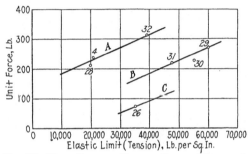

FIG. 72   UNIT FORCE — ELASTIC LIMIT (TENSION)

Tools: 4-deg. clearance, 30-deg. front rake.
Chip: 0.012 in. deep by 0.5 in. wide. Speed: 20 ft. per min.

Bar No. 32: 1.03 per cent carbon steel.	Bar No. 30: S.A.E. 2320 steel.
Bar No. 29: S.A.E. 2345 steel.	Bar No. 31: S.A.E. 3120 steel.
Bar No. 4: 0.15 per cent carbon steel.	Bar No. 28: S.A.E. 1035 steel.
Bar No. 8: Cast iron, unannealed.	Bar No. 8A. Cast iron, annealed.

Bar No. 26: Brass, rolled.

unit-force curve A is identical with that for Fig. 70. Curve B represents the elastic limits in tension for each material. The elastic limit for the 1.03 per cent carbon steel, 1, is 39,170 lb., which is lower than that for steels 2, 4, and 5, while its corresponding unit-force value is highest of all steels. The elastic limits of bars Nos. 3 and 6, both low-carbon steels, are relatively very low while the respective unit forces are even with those of the other steels. The elastic limit for brass, 7, is not altogether reliable, although it was determined very carefully. Curve D represents the elastic limits in compression for the materials and follows closely curve B, the elastic limits in tension. Curves C, E, and F represent the ultimate strength (maximum load divided by original area) in tension, compression, and shear, respectively (for test-bar shapes, see Appendix No. 3). These curves show high values for the alloy steels 2, 4, and 5, and relatively low values for the carbon steels 1, 3, and 6, and indicate no consistent relation to the cutting unit-force curve. The ultimate-strength curves E and C show a reduction for annealed and unannealed cast iron, which is not proportional to the reduced cutting-unit forces of curve A. The elastic limit and ultimate strength in tension, curves B and C are relatively high for brass, 7, however, which has the lowest cutting unit force. Figs. 70 and 71 indicate no consistent relation between the physical properties of the various metals and the unit forces. It does suggest, however, that it may be of advantage to divide the materials into four groups, namely, alloy steels, carbon steels, brasses, and cast irons for further study.

86   Figs. 72 to 80, inclusive, show the location on the coördinate axes of each material. The ordinates equal the unit force for the material for the 30-deg. tool and the abscissas equal the physical property in question. Fig. 72 shows the points plotted over the elastic limits in tension as abscissas. This shows bars Nos. 28, 4,

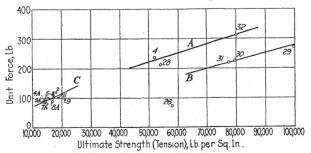

FIG. 73   UNIT FORCE — ULTIMATE STRENGTH (TENSION)

Tools: 4-deg. clearance, 30-deg. front rake.
Chip: 0.012 in. deep by 0.5 in. wide. Speed: 20 ft. per min.
(Material bar numbers same as in Fig. 72.)

and 32, which are all straight carbon steels, lying close to the straight line A, while bars Nos. 31, 30, and 29, all alloy steels, indicate line B, and the brass, bar No. 26, being isolated, might indicate the third line C. In Fig. 73, the points are plotted over the ultimate strengths in tension. Again the straight carbon steels, bars Nos. 4, 28, and 32, indicate the possibility of the curve A, while the alloy steels, bars Nos. 29, 30, and 31, indicate the possibility of curve B. The cast irons, bars Nos. 8 and 8A, and the brass, bar No. 26, do not seem to conform in any way with the values of the steels. Points for other cast-iron bars Nos. 1, 2, 4A, 5, 7A, and 9A all fall within half an inch to the left or above the point 8A, while bar No. 9 falls half an inch to the right of point 8A. This suggests a third line C which gives a relation between the unit cutting force for a 30-deg. front-rake tool for cast iron and its ultimate strength in tension. Fig. 74 shows these points again plotted over the elastic limit in compression and Fig. 75 the same for the ultimate strength in compression. In each case a curve A through the carbon steels is indicated, and also a curve B through the alloy steels. Again, points for all available data for cast iron are plotted. Except for bar No. 9, the points indicate that a line C expresses a relation between the unit force on the 30-deg. tool and the ultimate strength in compression of the cast iron tested. Curve C conforms well with curve B extended. Fig. 76 shows the points plotted over the ultimate strength in shear, and

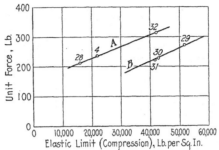

FIG. 74 UNIT FORCE — ELASTIC LIMIT (COMPRESSION)

Tools: 4-deg. clearance, 30-deg. front rake.
Chip: 0.012 in. deep by 0.5 in. wide. Speed: 20 ft. per min.
(Material bar numbers same as in Fig. 72.)

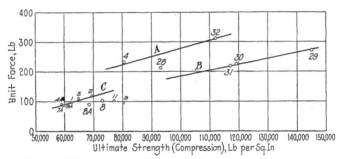

FIG. 75 UNIT FORCE — ULTIMATE STRENGTH (COMPRESSION)

Tools: 4-deg. clearance, 30-deg. front rake.
Chip: 0.012 in. deep by 0.5 in. wide. Speed: 20 ft. per min.
(Material bar numbers same as in Fig. 72.)

again there is an indication of a curve A for carbon steels and another curve B for the alloy steels. Points for all available data on cast iron are again plotted, and, while they fall close together, a curve C is indicated. Curve C might be shown to become a part of curve B extended, were more data available. The single point for brass, bar No. 26, seems to be quite independent.

87 Values of the unit forces for the various materials are plotted over the Brinell, the scleroscope, and the Rockwell hardness numbers in Figs. 77, 78, and 79, respectively. The Brinell

figures in Fig. 77 also seem to indicate a line A through the points 28, 4, and 32 for the straight carbon steels and another line B through the alloy steels 31, 30, and 29. The points for cast iron, bars Nos. 8 and 8A, and for brass, bar No. 26, seem to have the same unit-force values over a range of Brinell hardness numbers from 110 to 150, but the unit forces increase as the Brinell number increases above 150. The suggested curve C is of little value

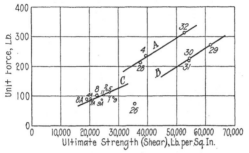

FIG. 76 UNIT FORCE — ULTIMATE STRENGTH (SHEAR)

Tools: 4-deg. clearance, 30-deg. front rake.
Chip: 0.012 in. deep by 0.5 in. wide. Speed: 20 ft. per min.
(Material bar numbers same as in Fig. 72.)

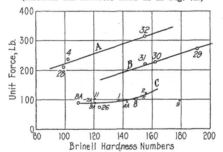

FIG. 77 UNIT FORCE — BRINELL HARDNESS NUMBERS

Tools: 4-deg. clearance, 30-deg. front rake.
Chip: 0.012 in. deep by 0.5 in. wide. Speed: 20 ft. per min.
(Material bar numbers same as in Fig. 72.)

because of the slight change in unit force for a wide range of Brinell reading. The scleroscope hardness numbers shown in Fig. 78 seem to group the steels together in one group and the brass and cast irons together in a second. There appears to be no relation which would predict the unit-force value for a given scleroscope hardness for the steels, but curve C through the cast-iron points is indicated. There appears to be no useful relation between the unit-force points plotted over Rockwell hardness numbers in Fig. 79, as indicated curve C is horizontal until a Rockwell number of 70 is reached, after which there is a small increase in the unit force for an increase in hardness. The points for the steels are grouped above the lower values of Rockwell hardness numbers, but in the higher unit-force values. The points for cast iron and brass extend over a wide range of Rockwell numbers from 57 to 88, but over a very limited range of unit force values which are of low magnitude.

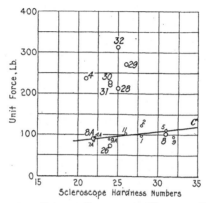

Fig. 78   Unit Force — Scleroscope Hardness Numbers

Tools: 4-deg. clearance, 30-deg. front rake.
Chip: 0.012 in. deep by 0.5 in. wide. Speed: 20 ft. per min.
(Material bar numbers same as in Fig. 72.)

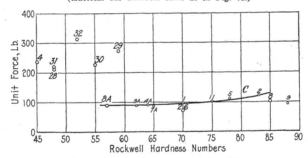

Fig. 79   Unit Force — Rockwell Hardness Numbers

Tools: 4-deg. clearance, 30-deg. front rake.
Chip: 0.012 in. deep by 0.5 in. wide. Speed: 20 ft. per min.
(Material bar numbers same as in Fig. 72.)

88   In Fig. 80 are shown plotted for each material the unit forces as ordinates over the reduction of area in per cent and elongation in two inches in per cent as abscissas. The reduction-of-area points cover the whole field from 32 per cent for brass, bar No. 26, to 70 per cent for S.A.E. 3120 steel, bar No. 31. The range in unit forces likewise covers the whole field and there appears to be no indication of the value of the unit force for a material having a known reduction of area. The percentage-of-elongation points

are well grouped together to the left of the figure. Line *A* has been drawn through the points for the alloy steels, bars No. 29, 30, and 31, and may represent a relation between the elongation and the unit cutting force. A similar relation is not apparent, however, for the carbon steels, as bar No. 28 falls too far below a line *B* drawn through points 32 and 4.

89 These deductions have been made in most cases for only three points for each line. Unquestionably, additional data for both carbon and alloy steels would permit more definite conclusions to be drawn. The unit forces are those for a 30-deg. front-rake tool. This in itself has an influence on the lines *A* and *B* of the various figures. If a 15-deg. front-rake tool had been selected instead of the 30-deg. tool, the unit-force curve would be higher throughout but the values for some materials would be relatively

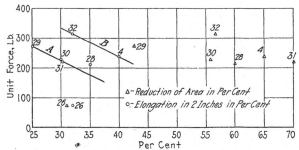

FIG. 80   UNIT FORCE — ELONGATION AND PERCENTAGE OF
REDUCTION OF AREA

Tools: 4-deg. clearance, 30-deg. front rake.
Chip: 0.012 in. deep by 0.5 in. wide. Speed: 20 ft. per min.
(Material bar numbers same as in Fig. 72.)

higher than those for others, as it was seen in Fig. 26 that the slope of the unit-force curve for front-rake angles is greater for some materials than others. For another heat-treatment of each of these bars, the unit force would be greater or less. This, too, might destroy the apparent relation between the physical properties and the cutting forces. The fact that the straight carbon steels seem to separate themselves from the alloy steels as far as the cutting properties are concerned is entirely new to the author and appears to be a problem of atomic structure of the materials.

SUMMARY

90 To enable one to compare the results obtained above, three sets of curves have been prepared as a function of cubic inches of metal removed per minute per horsepower. Fig. 81 shows these values for various front-rake angles in degrees as abscissas. For all metals, there is a distinct increase in the amount of metal removed per minute per horsepower as the front-rake angle is

increased, i.e., the cutting angle is reduced. For all of these curves the depth of cut is 0.012 in., the width of cut, 0.5 in., and the cutting speed, 20 ft. per min. It is observed from the figure that the 1.03 per cent carbon steel has the least amount of metal removed per horsepower-minute, increasing from 0.94 cu. in. for the 10-deg. front-rake tool to 1.52 cu. in. per min. for a 40-deg. front-rake tool. The unit forces are obtained from Fig. 26 and the cubic inches per minute per horsepower equals 396 divided by the unit force. These curves also show that the influence of the front-rake angle is less for S.A.E. 2345 steel than it is for the 0.15 per cent carbon steel, the latter showing an increase from

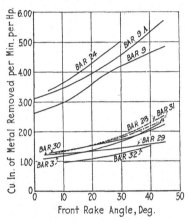

Front Rake Angle, Deg.

Fig. 81   Cubic Inches of Metal Removed per Minute per
Horsepower — Chip Width

Tools: 4-deg. clearance. Width of cut: 0.5 in. Depth of cut: 0.012 in.
Speed: 20 ft. per min.
(Bar No. 9, cast iron, unannealed; bar No. 9A, cast iron, annealed. Other material
bar numbers same as in Fig. 72.)

1.12 cu. in. per min. per horsepower for a 5-deg. front-rake tool to 2.37 cu. in. for a 45-deg. front-rake tool, while the corresponding figures for S.A.E. 2345 are 1.23 and 1.65 cu. in. The brass and cast iron are shown much higher on the scale than the steels.

91   Fig. 82 shows the cubic inches of metal removed per minute per horsepower plotted over the width of cut in inches for the various metals for the 30-deg. front-rake tool; for brass a 15-deg. tool was used. These curves fall off slightly for widths of cut between 0.1 and 0.3 in. The values for 0.15 per cent carbon steel (refer to Fig. 51 for unit forces) for a 0.2-in. width of cut is 1.77 cu. in. per min. per horsepower, and for a 0.7-in. width is 1.57 cu. in.; again the 1.03 per cent carbon steel is the lowest with about 1.5 cu. in. of metal removed per min. per horsepower for a 0.1-in. width, and 1.32 cu. in. for a 0.7-in. width of cut.

92   Fig. 83 shows a similar set of curves plotted over the depth of cut in inches for the 30-deg. front-rake tool; for brass a 15-deg. tool was used. In all cases, the amount of metal removed in cubic inches increases rapidly until a depth of cut of 0.010 in. is reached and then less rapidly until a depth of cut of about 0.025 in. is reached, after which the increase for the steels is less marked. Again, the lowest curve is for the 1.03 per cent carbon steel, which has a value of 1.06 cu. in. removed per min. per horsepower for 0.005-in. depth of cut and increases to 1.25 cu. in. for 0.010-in. depth of cut and to 1.73 cu. in. for 0.040-in. depth of cut. The 0.15 per cent carbon steel increases from 1.15 in. for a 0.0025-in. depth of cut to 1.86 cu. in. for a 0.025-in. depth of cut.

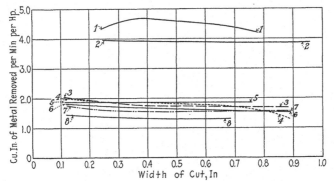

FIG. 82   CUBIC INCHES OF METAL REMOVED PER MINUTE PER
HORSEPOWER — CHIP WIDTH

Tools: 4-deg. clearance, 30-deg. front rake (except brass, 15-deg.)
Depth of cut: 0.012 in. Speed: 20 ft. per min.
Curve 1: Cast iron, annealed, bar No. 7A.
Curve 2: Brass, bar No. 34.
Curve 3: S.A.E. 2320 steel, bar No. 30.
Curve 4: S.A.E. 3120 steel, bar No. 31.
Curve 5: S.A.E. 1035 steel, bar No. 28.
Curve 6: 0.15 per cent carbon steel, bar No. 5.
Curve 7: S.A.E. 2345 steel, bar No. 29.
Curve 8: 1.03 per cent carbon steel, bar No. 32.

93   These curves indicate clearly the influence of the variables involved on the cutting efficiency of the tool. For these curves the cubic inches removed per horsepower-minute has been used as the ordinate, but some authors have used the inversed ratio, that is, the horsepower per cubic inch of metal removed per minute. These values are readily interchangeable, one being the reciprocal of the other.

CONCLUSIONS

94   The following conclusions are based only on the data collected by the author as presented above, and are not influenced by the work of other investigators. They are, however, compared

later with results of others. In reading these conclusions it should be kept in mind that this is an investigation of forces on the tool as a function of some one variable, rather than a study of tool endurance. A conclusion quite favorable from the standpoint of force on the tool may be decidedly unfavorable to the life of the tool. The conclusions follow:

1 The burred keenness of a newly ground tool edge disappears after the first two or three feet of cutting. The tool edge main-

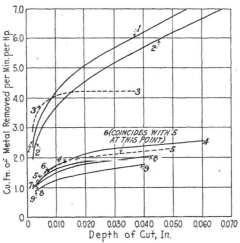

FIG. 83 CUBIC INCHES OF METAL REMOVED PER MINUTE PER HORSEPOWER — DEPTH OF CUT

Tools: 4-deg. clearance, 30-deg. front rake (except brass, 15-deg.)
Width of cut: 0.5 in. Speed: 20 ft. per min.
Curve 1: Cast iron, annealed, bar No. 8A.
Curve 2: Cast iron, unannealed, bar No. 8.
Curve 3: Brass, rolled, bar No. 26.
Curve 4: S.A.E. 1035 steel, bar No. 28.
Curve 5: S.A.E. 3120 steel, bar No. 31.
Curve 6: S.A.E. 2320 steel, bar No. 30.
Curve 7: 0.15 per cent carbon steel, bar No. 3.
Curve 8: S.A.E. 2345 steel, bar No. 29.
Curve 9: 1.03 per cent carbon steel, bar No. 32.

tains this new condition while cutting several hundred feet. When a second change occurs, the failure of the tool soon follows.

2 The force on the tool for a given material and size of cut is the same for a tool carefully heat-treated, and ground on any standard commercial abrasive wheel. A tool with a carefully honed cutting edge gives the same force as a tool ground on a coarse-grain wheel; also a slight rounding of the edge with a honing stone causes no increase in the force on the tool, other conditions being constant. The thicker the chip, the blunter the cutting edge may be without producing a noticeable increase in the force on the tool (see Fig. 8).

3   A variation of the clearance angle between 1 and 10 deg. has no influence on the force on the tool in the direction of the cut, all other conditions being constant (see Fig. 16).

4   The force on an end cutting tool for all materials is reduced in direct proportion to the increase of the front rake angle, until a certain limiting angle is reached. Beyond this limiting angle, which appears to be between 20 and 30 deg. for brass, 25 to 30 deg. for cast iron, and above 45 deg. for steel, the influence of the front rake angle is less pronounced, i.e., for the same increase in front-rake angle, the unit force is reduced a smaller amount. For front-rake angles above 60 deg. that is, for cutting angles less than 30 deg., the total pressure on the tool in the direction of cut is increased (see Fig. 25). For some materials, however, such as S.A.E. 2345, the force on the tool is influenced less for a given increase in front-rake angle than others (see Fig. 26).

5   The introduction of side rake on an end-cutting tool has no influence on the unit or total force on a tool in removing a given chip. For a tool having front and side rake, cleaner cutting was noted than for the 0-deg. front-rake tool with various angles of side rake (see Fig. 34). For values of side rake of 60 deg. or greater on the 0-deg. front-rake tools, an increase in the force on the tool in the direction of cut occurs (see Figs. 35 and 36).

6   For the same material, tool, and depth of cut, the total force on the tool increases with an increase in width of cut, but at a greater rate. The total force on the tool in the direction of the cut plotted over the width of cut gives a curve slightly concave upward, passing through the origin; the curvature increases as the width approaches zero (see Fig. 50). In other words, all other conditions remaining constant, a chip less than 0.25 in. in width is removed more efficiently than a chip wider than 0.25 in. This applies for rectangular chips removed with an end cutting tool (see Fig. 51).

7   For the same material, tool, and width of cut, all other conditions being constant, the total force on the tool increases as the depth of cut increases, but at a slower rate. The total force on the tool in the direction of cut plotted over the depth of cut gives a curve concave downward and passing through the origin; the curvature increases as the depth approaches zero (see Fig. 68). The force on the tool per 0.001 sq. in. of cross-sectional area of chip is much higher for a thin chip than for a thick chip; this difference is greatest, however, for increments in the small depths of cut. When the depth of cut is in excess of 0.030 in., the unit-force curve becomes almost horizontal. The unit force increases rapidly as the depth of the cut is reduced below 0.030 in. This applies for rectangular chips removed with an end cutting tool (see Fig. 69).

8   When the unit cutting forces are plotted against the various physical properties, no solution is indicated when all metals are

taken as a group. For certain physical properties the materials divide themselves into three groups, namely, straight carbon steels, alloy steels, and cast iron, for each of which a relation is indicated. For the straight carbon steels it appears that the elastic limit in tension and compression, the ultimate strength in tension and shear, and the Brinell hardness are functions of the unit force on the tool. For the alloy steels the elastic limit in tension and compression, the ultimate strength in tension, compression, and shear, the Brinell hardness, and the percentage of elongation in two inches, each indicates a relation to the unit force. For cast iron a similar relation is suggested for the ultimate strength in tension, compression, and shear, and the scleroscope hardness number (see Figs. 70 to 80 incl.).

COMPARISON AND DISCUSSION OF RESULTS

95 Investigations in metal cutting have been made under so many different conditions, by so many individuals, and over such a long period of time that it is with considerable circumspection that the author attempts any summarization and comparison of results. References to a few papers, however, which deal directly with the problems of this paper are made. The author believes that a thorough digest and summarization of existing data is one of the most important steps for the benefit of the future of the metal-cutting art. The theories should be substantiated by results from practice. Such a summarization would doubtlessly indicate that more data are needed on some subjects to confirm, explain, or augment those already available, and certainly would bring to light much that has never received sufficient publicity, and that which has been forgotten or never published in this country. If possible, a theory of metal cutting which underlies all types of cutting should be developed, as certainly there is something in common in the removal of a chip by milling, drilling, turning, etc. All this is a tremendous problem and should be undertaken in a big way.

96 The terms applicable to metal cutting should be standardized, and the size and shape of test specimens for deriving physical properties of the metals cut should be simplified and standardized.

97 Cutting fluids should receive a due portion of study, as they are a necessary part of cutting metals. We know that cutting fluids have a great influence, but the best cutting fluid (lubricant and coolant) for one job is hopelessly inadequate for the next; why?

Comparison: Sharpness of tool — Problem (a)

98 Just what happens when a tool cuts has been very clearly shown by E. G. Coker[1] by passing polarized light through nitro-

---

[1] Ref. D-35, p. 575.

cellulose when being cut by a steel or glass tool of the end-cutting type when milling, turning, and planing. The resultant color effect caused by the strained material indicated definitely the strains set up in the tool and the material. He observed that a tool ground in the ordinary way (later referred to as badly ground) with an angle of 45 to 60 deg. produced an action quite different from one with a similar edge but finished on a stone as perfectly as possible. In the former case the cutting action was irregular and imperfect, material being broken away from the piece by a wedge action in which apparently the shaving was bent away from the disk being cut by the upper face of the tool. The finished tool gave a continuous, smooth-flowing, clean-cut chip. Coker reports further, "the edge of the disk is left in a ragged condition and somewhat wavy in outline, so much so, in fact, that in some cases the main shaving broken off in this way is accompanied by a very thin shaving quite separate from the main one, and pared off by a true cutting process. That it is possible to obtain a somewhat similar effect, in the turning of cast iron is well known, only in this case the cast-iron chippings are accompanied by a fine powder of cast iron falling from the cutting edge of the tool." Whether there is a difference in the forces acting on the tool in the two cases is not stated by Coker. Fig. 8 of the present paper would indicate that the force should be the same unless nitrocellulose has cutting properties different from cast iron and steel, also unless the cutting tool is considerably duller than the one referred to as being "ground in the ordinary way." Fig. 9 shows that the finish for the honed tool is best of all, which agrees with Coker's results, but does not indicate that there should be such a difference in chip distortion due to the increased dullness. With tools definitely dulled, such excessive distortion was not observed as is shown in Fig. 9. The wedge action referred to by Coker was observed by the author, and for a given degree of tool sharpness occurs for deep cuts but not for fine cuts. A heavy chip shows the slippage lines due, evidently, to the shearing of the metal by the wedge, a very light or finishing cut is perfectly smooth on the bottom of the chip, which appears clean cut and burnished rather than torn. The inner strip of the chip referred to by Coker is undoubtedly that part of the material removed by actual cutting action of the point, whereas the major portion of the chip is removed by a combination of tear and shear. Coker also observed that the total work done was a linear function of the speed. The author noticed for many different conditions that the force on the tool was the same for speeds of 20 and 35 ft. per min., the two available cutting speeds of the planer.

99 Crompton,[1] reported that with heavier cuts the keen-honed edge lasted when cutting steel for only a few feet of cutting, after

[1] Ref. D-35, pp. 597-598.

which the edge became slightly rounded, in which condition it remained for some time. Owing to the scouring action of the chip on the upper face of the tool, concavities were formed which, as they deepened, advanced toward the cutting edge and eventually met it in such a manner that the included angle between the curved top surface and the front face of the tool was lessened, the cutting efficiency being thereby much increased. This condition continued for a long time and eventually the scouring action of the chip, together with the wear on the leading face, so reduced the lip angle of the tool that it broke away. He also reported that with a slightly blunted tool, the chip, in all cases of steel, consisted of a very thin inner unbroken layer which was continuous, on the top of which the rest of the chip was piled in a series of slabs. Dean Ripper [1] found that the wear of the cutting edge was continuous while cutting, even though a false lip or wedge of cut material protected it from most of the work of cutting. Many curves are given showing that the relation between the units of wear of the edge and time of cutting for a constant cutting speed is a curve passing through the origin, slightly concave downward. The curve is steepest for the highest speeds.

100   Fred A. Parsons [2] reported that " the condition of the cutter has a pronounced effect on the amount of power required for removing metal. Tests showing the variations for sharp and dull milling cutters cannot be expected to show uniform results, as there is no standard for dullness. However, a number of tests show that the power may be expected to increase as much as 40 per cent or more before the appearance of the cut warns the operator that it is time to resharpen in the case of milling cutters." Parsons further states that other tools such as drills and lathe tools also cause an increase in power as they become dull. The author believes that the increase in power consumption for continuous-cutting tools as they become dulled is not so marked as in the case of milling cutters. Milling is peculiar in that each tooth removes only a small chip, which starts with zero thickness and increases to a maximum; before the tool starts to cut the tooth exerts a rubbing action over the work, which is certainly a function of the keenness of the edge. In continuous turning, a false cutting edge is built up on the tool from the material being cut, which partially protects the cutting edge and also relieves the cutting edge of most of the work. In milling, most of the chip would be removed before this false edge is built up to assist the cutting and protect the cutting edge. Coker's and Crompton's results agree with the observations of the author. Ripper's statement that tool wear was uniformly continuous does not agree with Crompton or the author. Parson's report on milling cutters

[1] Ref. D-16, p. 1074.
[2] Ref. C-3, par. 50.

does not agree. The variance seems plausible, however, because of the rubbing action of each tooth of the milling cutter before it starts to cut.

Comparison: Clearance Angle of Tool — Problem (b)

101   F. W. Taylor[1] stated: " in seeking for the proper clearance angles for tools, we have as yet been unable to devise any type of experiment which would demonstrate in a clear-cut manner which clearance angle is the best." Taylor reported further that " on the one hand, it is evident that the larger the clearance angle, the greater will be the ease with which the tool can be fed into its work, the first action of the tool when brought into contact with the forging being that of forcing the line of cutting edge into the material to be cut. On the other hand, every increase in the clearance angle takes off an equal amount from the lip angle and therefore subjects the tool to a greater tendency to crumble or spall away at the cutting edge. It must be remembered also that the tool travels in a spiral path around the work which it is cutting in the lathe, and that the angle of this path with a perpendicular line in the case of coarse feeds taken upon small diameters of work, becomes of distinctly appreciable size. In all cases, therefore, the clearance angle adopted for standard shop tools must be sufficiently large to avoid all possibility from this source of rubbing the flank of the tool against the spiral flank of the forging. The clearance angles for roughing tools in common use vary between 4 and 12 deg. We have had experience on a large scale in different shops with tools carefully ground with clearance angles of 5, 6, and 8 deg. In the case of one large machine shop which had used clearance angles ground to 8 deg. through a term of years, they finally adopted the 6-deg. clearance angle with satisfaction. For many years past our experiments have all been made with the 6-deg. clearance angle, and this has been demonstrated to be amply large for our various experiments. On the other hand, a 5-deg. clearance angle in practical use in a large shop has appeared to us through long-continued observations to grind away the flank of the tool just below the cutting edge rather more rapidly than the 6-deg. angle. We have therefore adopted the 6-deg. clearance angle as our standard." Taylor points out further that " a clearance angle of from 9 to 12 deg. should be used in shops in which each machinist grinds his own tools." In straight-line cutting such as planer and shaper work, the 4-deg. clearance angle, as used by the author, is considered ample. For broaching, another form of straight-line cutting, even smaller angles are used. H. F. Donaldson[2] states that the clear-

[1] Ref. D-8, par. 336 and 337.
[2] Ref. D-6, p. 7.

ance angle should under no circumstances be less than 3 deg., but should be greater as the diameter of the work in the lathe increases. The influence of the diameter of the work was discussed by Orcutt, on page 29, who believed the statement to be an error. Ashton, on page 38, agreed with Donaldson in that "for a small radius of curvature the surface of the work was less in contact with the tool than with the large radius, that is to say, that it could stand a flatter and s ,uarer angle." In the Manchester experiments of 1922, reported on by Dempster Smith,[1] a 6-deg. clearance angle was used. Airey and Oxford[2] reported that "excessive clearance does not reduce the cutting force, does not appreciably affect the life of the tool in normal wear, but does tend to chatter in action; also it weakens the cross-section near the cutting edge and increases liability to snip. Clearance should therefore be kept to a minimum. Three degrees was found to be ample, but 5 deg. was usually used. In many tool cribs it might be desirable to standardize at 8 deg. due to the possibility of an error of a few degrees." The results of the experimenters mentioned above are in accordance with the author's. Taylor's 6-deg. clearance angle for turning was based on wear or tool life with no reference to the force on the tool involved.

Comparison: Front-Rake Angle — Problem (c)

102   Robert H. Smith,[3] before 1880, found that in turning with a round-nose tool, a side rake angle — with practically no front rake (for turning this is equivalent to the author's front rake angle) — of 24 deg. for cutting cast iron and $46\frac{1}{2}$ deg. for steel and wrought iron gave very satisfactory results. The tools lasted admirably. He determined that the smaller the cutting angle of the tool edge, the smaller was the horsepower required for cutting; but the angle actually required depended upon the necessity of avoiding, first, too frequent breakage, and second, overheating and consequent loss of temper at the edges. He further reported: "it was well known that with a large amount of top rake there was risk of 'digging-in.' If the depth of cut were small, the work would push the tool outwards from the center, a push which the tool clamp must resist. If the cut were deep, it would draw the tool in toward the center, to prevent which the tool clamp must exert an outward pull. For each top rake and for each quantity of material cut — perhaps varying slightly with the cutting speed — there was one definite depth of cut, to maintain which steadily the tool clamp needed to exert no radial force whatever. Similarly with the transverse feed."

[1] Ref. D-33,. p. 9.
[2] Ref. C-1, par. 75.
[3] Ref. D-6, p. 52.

103 Airey and Oxford[1] report for milling that " from a power-
consumption standpoint, rake is increasingly beneficial as it
becomes greater. The advantage of rake, though, does not increase
at so fast a rate after about 15 deg. is past. Further, this is influenced by the kind of material being cut. On the other hand, the
life of the cutter might be influenced disadvantageously by excessive rake." They show in Fig. 4 that for milling the metal removed
in cubic inches per horsepower-minute is increased due to increased front rake, but not as a direct function of the rake angle.
Calling the production of the zero-degree-rake tool over a certain
region of chip weight as 100, then the productions for the 10-, 20-,
and 30-deg. tools are, for

> Cast iron 133,   138.4, and 142.1
> Bronze .... 113.2, 123,   and 118
> Machine steel .... 118.7, 157.2, and 172
> Carbon tool steel.......... 106,   112.2, and 112

104 The Manchester experiments[2] show that the vertical force
on the tool when turning increases almost directly as the cutting
angle in the plane of shaving increases. This true cutting angle
is actually a function of the front and side rake of the tool and
corresponds quite favorably with the front rake of the author's
experiments, although the chip formation itself is somewhat different, inasmuch as a round-nosed tool was used in the Manchester
experiments, while a straight-edged, end-cutting tool was used by
the author. Smith[3] shows that the cutting speed for a definite
tool life for a number of steels is maximum when the cutting
angle in the plane of shaving is between 70 and 75 deg. and that
the intensity of pressure on the top surface of the tool increases
with the cutting angle (reduced rake) resulting in a higher rate
of wear, but at the same time the amount of metal to be worn
away before the cutting edge is reached increases.

105 Klopstock[4] shows that the power consumption is considerably reduced for small contained angles, that is, for large
front-rake angles, but that there is little difference in power consumption for large contained angles. This work refers to Taylor's
statement,[5] which says: " The writer believes that it would be
profitable to experiment with more acute lip angles than 61 deg.
in cutting dead-soft steel." It may be that with this extremely
soft steel, still higher cutting speeds could be obtained with more
acute angles, in which case it would be advisable, of course, to
make special tools for cutting this quality of metal, in shops where

[1] Ref. C-1, p. 28.
[2] Ref. D-33, p. 75.
[3] Ref. D-33, p. 51.
[4] Ref. D-72, p. 14.
[5] Ref. D-8, p. 91.

large amounts of it are used. Klopstock's interpretation of Taylor's statement as quoted by DeLeeuw does not agree with the results shown in this paper. The author believes, however, that Taylor's statement is correct.

106   The results of tests conducted by Parsons [1] seem to show that for milling cutters, " the power required varies directly as the ratio of the conversed sine of the rake angle, or the cubic inches of metal removed per tool horsepower varies inversely in the same way." Parson's tests carried the front rake angle only to 12 deg., however.

107   Rosenhain and Sturney report [2] that the average depth of cut is very much greater than that intended (0.010 or 0.015 in. as a constant) when small top-rake angles are used, but decreases rapidly as the top-rake (front-rake) angle increases until, with an angle of about 15 deg., when cutting mild steel, the average approximates closely to the intended depth of cut. The conclusions of the experimenters referred to above, with the exception of Klopstock, are in agreement with those of the author. Their conclusions, however, are based on the more limited practical cutting range while the author's are based on the whole range that the tool could be made to cut.

Comparison: Side Rake — Problem (d)

108   Airey and Oxford [3] state that " it is difficult to understand the growth of the belief that spiral cutters are more efficient, because abstract analysis leads to the conclusion that spiral cutters must be inferior to straight cutters. By ' inferior ' the meaning intended is in reference to power efficiency only. The use of a spiral angle is heartily recommended, for it results in continuity of action, tends to avoid chatter, and keeps the driving power more smoothly constant. The smallest angle consistent with smoothness of action should be used."

109   For milling, the side-rake angle is comparable with the side-rake angle of the tools used by the author. In turning, however, the side-rake angle is really the front-rake angle as used by the writer, so that it is the side-rake angle in turning that is of most importance as far as power efficiency is concerned. The front-rake angle in turning is of value, however, as it affects the surfacing force on the tool and the direction of flow of chip.

110   Parsons [4] states that " the effect of adding a spiral angle to a milling cutter, as far as power efficiency is concerned, seems to be confined to reducing the bumping action of the cut and

[1] Ref. C-3, par. 42.
[2] Ref. D-67, p. 27.
[3] Ref. C-1, p. 19.
[4] Ref. C-3, par. 47.

thereby somewhat reducing the maximum power required. From the standpoint of power required, this effect is not important." " The spiral angle is of considerable importance, however, since a large spiral angle enables fewer teeth to be used in a spiral or slabbing cutter." The only available results on the influence of side rake alone, those for milling, as given above, agree with the author's in that side rake alone has little influence on the power consumption but does permit some desirable practical features.

Comparison: Chip Width and Depth of Cut — Problem (e)

111 F. W. Taylor[1] points out that for depths of cut less than $\frac{3}{16}$ in., an error is likely to be made in maintaining a uniform depth of cut, which becomes so large a percentage of the total depth as materially to effect the accuracy of the experiment. He states further that " this is true to such an extent that as a result of our experience we should consider practically worthless for determining laws all cuts that are as shallow as $\frac{1}{16}$ in. and we should much prefer a $\frac{3}{16}$-in. depth to $\frac{1}{8}$-in. depth." A cut $\frac{3}{16}$ in. in depth by $\frac{1}{16}$ in. feed is recommended by him. The author has referred, in connection with his data, to the irregularities or fluctuations when depths of cut less than 0.006 in. were taken. Taylor further states that " a feed which is finer than $\frac{1}{16}$ in. is also undesirable for experiments because any trifling imperfection or flaw at or near the cutting edge of the tool will more seriously affect the results with a smaller feed than this and also small hard spots or other blemishes in the metal that is being tested have a much worse effect on the tools with a fine than with a coarse feed." It was for this reason that the author felt it necessary to carry out the test on problem (a), involving the effect of the force on a tool as a function of its sharpness for various depths of cut, in which it was shown that a dulled tool increases the cutting force on thin chips but not on thick ones (see Fig. 8). Taylor further reports (par. 291) that " it is the thickness of shaving them which must be first considered, as this element has more effect upon the design of our standard tools, and in fact upon the whole problem of cutting metals, than any other single item which is completely under the control of those who are managing a shop." Taylor concluded that the pressure in pounds per square inch of chip area increases as the chip becomes thinner. This agrees with the early experiments conducted in Germany but disagrees with those of Nicolson[2] who concluded that the pressure on the tool per square inch of cross-sectional area of chip was a direct function of the area of the chip and was the same whether light or heavy cuts were taken, and that it did not depend upon

---

[1] Ref. D-8, p. 66.
[2] Ref. D-7.

either the thickness of the feed or the depth of cut. As to the chip thickness affecting the cutting speed, Taylor wrote: " to make it more apparent that the element affecting the cutting speed the most is the thickness of the shaving, the writer would call attention to the fact that dividing the thickness of the shaving by 3 increases the cutting speed in the ratio of 1 to 1.8, while dividing the depth of cut by 3 only increases the cutting speed in the ratio of 1 to 1.27."

112   Smith[1] reported from the Manchester experiments, for the vertical force on a tool when turning in a lathe, that " the results obtained in these trials are plotted on a base of depth of cut in Fig. 44, and it will be observed that for each traverse the vertical force increases with the depth of cut according to a linear law. It will also be seen that for a given depth of cut this force does not vary directly as the traverse, but is proportionately greater for finer traverses." Smith showed in his Fig. 30 that for a constant depth of cut the cutting speed for a definite tool life was maximum for fine traverses, but was reduced very rapidly for small increases in traverse and gradually turned from a steep curve for small traverses to nearly horizontal for heavy traverses. Also, for a constant traverse with variable depth of cut, the cutting speed for a definite tool life was also shown to be maximum for the smallest depth of cut, reducing rapidly for the first increase in depth and less rapidly with additional increases until the curve becomes almost horizontal for depths between $\frac{3}{16}$ and $\frac{3}{8}$ in. Airey and Oxford[2] state, " it has been demonstrated earlier that the force required to remove metal (by milling) does not increase in proportion to the chip thickness. It follows directly from this that as feed is increased, the force will not increase in proportion."

113   Klopstock[3] states, " It will be seen that a chip of 1 mm. depth of cut and 10 mm. feed will require only about one-half the actual power (delivered to machine) as compared with a chip of 10 mm. cut and 1 mm. feed." For these same cuts it is seen from his Fig. 13 that the actual power absorbed by the first chip is only one-third or one-fourth of that absorbed by the second chip. In Klopstock's Fig. 18 is shown the total and unit vertical pressures on the tool for wrought iron and cast iron, the curves of which compare favorably with those of the author (Figs. 60 to 64). In his Figs. 19 and 20 are shown the total force plotted against the cross-sectional area of the chip as abscissas for various experimenters such as Taylor, Nicolson, Ripper, and Klopstock. These curves, while plotted on the cross-sectional area of chip as a basis, fall within the limits of the results of the author. It is observed, however, that if the chip area is made up with a constant

[1] Ref. D-33, p. 80.
[2] Ref. C-1, par. 51.
[3] Ref. D-72, p. 12.

width, the area being increased by an increase in depth, a curve slightly above those shown is obtained, which is concave downward. As the area of chip is increased, by increasing the width, that is for constant depth, a curve below those shown, which is concave upward, is obtained. (See author's Figs. 43 and 62 for nickel-chrome steel for example.)

114 Ripper[1] concluded (p. 1092, No. 6) that for carbon-steel tools, with which cuts of small area only are possible, the influence of the depth of cut upon the cutting speed for certain tool life is exactly the same as that of the feed, also (p. 1118, No. 5), " if the area of cut is kept constant, a higher cutting speed for given tool life is obtainable when the cut is deep and the feed fine, than when the cut is shallow and the feed coarse." This effect is undoubtedly due to greater length of cutting edge to dissipate the heat than because of smaller forces involved.

115 Parsons[2] states that " a test run first with several narrow cutters set up with all the teeth on a line, and again run with the teeth staggered (to give the effect of a spiral cut) showed that the power reduced in the ratio of 1.42 to 1.27." It seems very probable to the author that this reduction in power is due not to the effect of a spiral, but to the effect of narrower widths of chip as is pointed out above.

116 Rosenhain and Sturney[3] show that for a given front rake angle, the average depth of cut for an end-cutting tool is greater than the intended depth of cut as a result of the tearing of the chip ahead of the cutting edge but slightly deeper into the metal being cut. This condition is increased as the depth of cut is increased from 0.01 to 0.035 in. Most of the published results on the relation of cutting force to chip depth agree in general with those of the author. The author's method, however, permits a careful study of the single variable and shows the limiting values of both width of cut and depth of cut under which the unit force is increased as the depth is reduced or decreased as the width is reduced.

Comparison: Physical Properties — Problem (f)

117 Many experimenters have compared physical properties of materials cut with the force on the cutting tool. The difficulty in comparing such results, however, is that the conditions have not been properly standardized.

118 Machinability is a term which during the last few years has gained considerable prominence, but, like the hardness of metals, is not clearly defined. The author for the sake of

[1] Ref. D-16, p. 1092 and 1118.
[2] Ref. C-3, par. 48.
[3] Ref. D-67, p. 27.

comparison selected the force on a 30-deg. front-rake tool for a chip 0.012 in. deep and 0.5 in. wide as representing the machinability of the material. A machinability tester, described in Airey and Oxford's paper on The Art of Milling, has a cutter functioning as a single tooth of a milling cutter, which takes a width of cut of $\frac{1}{8}$ in., a radius of cut of $1\frac{3}{4}$ in., with a feed per chip of 0.004, 0.008, 0.012, and 0.016 in. for a constant depth of cut. This machinability tester measured the energy in foot-pounds required to remove a chip of a given metal.

119   Other types of machinability testers of the drill type which have been used by Bauer,[1] Keep,[2] Leyde,[3] Reininger,[4] Kurrin, and Kessner[5] use some factor of the penetration of the drill as the indication of machinability of a metal. Grossmann[6] states that "the drill test can give only comparative, not absolute, values for machinability, since the angle a $\left(\tan a = \dfrac{\text{revolutions}}{\text{depth}} \right)$, Fig. 255, will depend not only on the material under test, but also on the material of the tool, the cutting angles, the pressure P, the rate of rotation and the progressive dulling of the tool."

120   M. A. Grossmann in his Physical Metallography,[6] p. 326, gives a very complete discussion of the relation of machinability to hardness and workability. He states that "by workability is understood the capacity of a material for undergoing plastic deformation without rupture. The elongation in a tension test, the extent of bending before fracture in the notch-impact and various cold-bend tests, and the extent of deformation in the other mechanical tests all represent this quality of workability." After explaining Thieme's conception of chip removal, which is a pressure process, he adds (p. 328), "There comes into question not only the resistance to penetration, but also the resistance to the pressing aside of the chip element," and concludes with, "we arrive thus at the law that machinability depends on both hardness (e.g., ball hardness) and workability, the difficulty of machining increasing with an increase in either property." It is explained that the test may be carried out on a planer, lathe, or drill press, all factors constant so that the travel of the tool for a given force on it in a specific time, is the measure of machinability. A case is given in Fig. 256 to show the machinability as determined by means of the Kessner drill apparatus and the ball hardness of a copper (2)-zinc (1) alloy with various amounts of lead from zero to

[1] Ref. D-2.
[2] *Iron Age*, 1899, p. 9, and 1900, p. 16.
[3] *Z. d. Ing.*, 1904, p. 169.
[4] Giessereizeitung 1904, pp. 217 and 627.
[5] Ref. D-60.
[6] Ref. D-86, pp. 326 and 330.

11 per cent. The machinability (drill penetration) curve is a smooth curve concave downward, increasing in height rapidly as lead is first added, but gradually becoming nearly horizontal when the lead content is 11 per cent. The ball-hardness curve, on the other hand, is practically the same for the 0 and 11 per cent contents, but bends upward to a maximum for a lead content of 2 per cent.

121   Klopstock[1] shows in his Fig. 23 Brinell-hardness curves and chip-pressure curves for seven materials, including Kurrin's drilling test. He states that "it appears that the intersecting points of the hardness curves and chip-pressure curves will be found near two approximately parallel straight lines; one located considerably below the other. The upper line connects the inter-secting point of curves relating to materials forming continuous chips such as steel, wrought iron, copper, etc., while the lower line connects the intersecting points of curves relating to materials such as cast iron and brass." "These observations permit the determination of chip pressure and, therefore, power requirements in the turning and planing of materials of which the Brinell hardness characteristics are known."

122   The Erichsen test supplemented with a photomicrograph was selected as the best indication of drawability (workability) of the ten samples of sheet metal recently submitted by the Boston Sheet Metal Company to the Pennsylvania State College[2] for tests

123   The following is taken from a letter to the author from E. G. Herbert of Manchester, England: "I think it is quite well established that there is no direct relationship between the hard-ness of metals and their resistance to cutting, and it has been my endeavor in 'The Pendulum' and elsewhere to show why such a relationship cannot exist. The principal reasons are, accord-ing to my experience, two:

"(1) When a metal is cut it is work-hardened in the process of cutting, and its resistance to cutting depends far more on its work-hardening properties, that is on the hardness induced in it by the tool, than on its original hardness. As a familiar example, manganese steel is shown by the Brinell test to be soft. This is confirmed by the pendulum time test which gives its hardness 24, the hardness of ordinary mild steel being 20. But manganese steel cannot be cut, and the reason is immediately made apparent by the pendulum 'work-hardening test' which gives its original scale hardness 14 and its scale hardness after being rolled with the pendulum ball 80 to 90, i.e., equal to the hardness of hardened tool steel. All other metals are similarly hardened by cutting tools, but in very different degrees, and their resistance to cutting must therefore depend on their 'work-hardening capacity.'

[1] Ref. D-72.
[2] *Mechanical Engineering,* Feb. 1926.

" (2) Metals are heated in cutting, and their resistance to cutting must therefore depend on their properties when in a heated state — not cold. The hardness of steels and other metals does not generally change much within the ordinary range of cutting temperatures, but their work-hardening properties do change in a remarkable manner, almost disappearing in many cases at temperatures (in mild steel) between 100 and 150 deg. cent. Any study of machinability which fails to take these facts into account must, in my opinion, lead to negative results."

124   Herbert also writes,[1] " The depression in the work-hardening curve coincides with the free-cutting range of temperatures. Metal cut within this range is only slightly hardened by the tool with the following results:

1   The metal cuts freely, leaving a smooth finish
2   Less heat is generated
3   A flowing helical chip is produced
4   A bright Whitaker ring is formed.

These results confirm the observations of Stanton and Dempster Smith (Bulletin of the Institution of Mechanical Engineers, No. 2, 1925) that the vertical force on the tool falls to a minimum at certain cutting speeds and also at higher speeds. Stanton found that this effect disappeared when the steel was normalized." It is not possible to compare the relations between physical properties of materials and their machinability as outlined above with those of the author because of the difference in methods or standards used. It appears obvious, however, that there is no satisfactory method extant which is wholly reliable for all metals.

[1] Ref. D-88 and D-90, p. 364.

APPENDICES

APPENDIX NO. 1-A

TOOL LIST

Manufacturer of steel, ...
Size of tool, 2¼ x 1¼ x 12 in.
Grade of tool, Special No. 14, p. 27, Cat. No. 23
Chemical composition, C, 1.05%; Mn, 0.22%; Si, 0.10%; S, 0.018%; P, 0.017%

Tool No.	Angles of tool form, deg.			Forged or machined	Anneal-ing	Forg-ing	Harden-ing	Temper-ing	Hardness, Rockwell	Remarks
	Clear-ance	Front rake	Side rake							
A-1	6	0	0	M	1380–1440	....	1400–1450	450	61	First end ....
A-1ᵃ	8	0	0	M	....	....	....	....	..	Rehardened and reground....
A-2	2	0	0	M	....	....	1430	450	60	Reground ....
A-3	3½	0	0	M	....	....	....	....	..	Reground ....
A-3ᵃ	10	0	0		....	....	....	....	..	Reground ....
A-4	6	0	0	M	....	....	....	....	..	
A-5	4	5	0	F	....	....	1500	440	60	Hardened and reground....
AA¹-6	4	0	45	M	....	....	1475	425	61	Other end ....

¹ Double letter indicates other end of bar.

APPENDIX NO. 1-B

TOOL SUMMARY SHEET SHOWING PROBLEMS USING THE TOOL

Tool No.	Angles of tool form			Problem tool was used on
	Clear-ance	Front rake	Side rake	
A-1	6	0	0	*b*
1*a*	8	0	0	*b*
2	2	0	0	*b*
3	3½	0	0	*b, c*
3*a*	10	0	0	*b*
4	6	0	0	*b*
5	4	5	0	*b, c*
AA-6[1]	4[2]	0	45	*d*
B-1	6	10	0	*b*
1*a*	8	10	0	*b*
2	2	10	0	*b*
2*a*	10	10	0	*b*
3	4	10	0	*a*
4	4	10	0	*a, b*
5	4	10	0	*a*
6	4	10	0	*a*
7	4	10	0	*a*
8	4	10	0	*a*
9	4	10	0	*a*
C-1	6	20	0	*b*
1*a*	8	20	0	*b*
2	2	20	0	*b*
3	3¾	20	0	*b*
3*a*	10	20	0	*b*
4	6	20	..	*b*
5	4	20	0	*c*
D-1	6	30	0	*b*
1*a*	8	30	0	*b*
2	2	30	0	*b*
2*a*	10	30	0	*b*
3	4	30	0	*b*
DD-4[1]	4	0	30	*d*
5[1]	4[3] at rt. angle to edge	0	30	*d*
E-1	6	40	0	*b*
1*a*	8	40	0	*b*
2	2	40	0	*b*
3	3⅔	40	0	*b*
3*a*	10	40	0	*b*
4	6	40	0	*b*
5	4	20	0	*c*
EE-6	4	0	0	*d*
F-1	..	20	0	..
2	4	20	0	*a*
3	4	20	0	*a*
4	4	20	0	*a*
5	4	20	0	*a*
6	4	20	0	*a*
7	4	20	0	*a*
8	4	20	0	*a*
9	4	20	0	*a, c*
G-1	..	50	0	..
2	4	30	0	*a, c, d*
3	4	30	0	*a*
4	4	30	0	*a*
5	4	30	0	*a*
6	4	30	0	*a*
7	4	30	0	*a*
8	4	30	0	*a*
9	4	30	0	*a, c, d*

[1] Opposite end of bar.
[2] Angle measured in plane of motion.
[3] Angle measured in plane normal to cutting edge.

TOOL SUMMARY SHEET—Continued

Tool No.	Angles of tool form			Problem tool was used on
	Clearance	Front rake	Side rake	
H-1	6	0	0	b
2	6	10	0	b
3	4	0	0	c
4	4	0	20	d
HH-5	4	0	60	d
I-1	6	40	0	..
2	6	30	0	..
3	6	30	0	..
II-4	4	0	75	d
5	4 at rt. angles to edge	0	75	d
J-1	4	15	0	c
K-1	4	25	0	c
KK-2	4	0	10	d
L-1	4	35	0	c
M-1	4	40	0	c
N-1	4	45	0	c
O-1	4	30	30	d
P-1	4	30	20	d
R-1	4	30	10	d
S-1	4	60	0	c
T-1	2	75	0	c

APPENDIX NO. 2-A

MATERIAL RECORD SHEET FOR EACH BAR OF MATERIAL CUT

Bar Number ; machine steel 1
Material, O. H. machinery steel (one of a six-bar order)
Manufacturer, ... Steel Company
Chemical composition ; C, 0.15 ; Mn, 0.24 ; Si, 0.13 ; S, 0.029 ; P, 0.014
Condition at factory, fully annealed, Brinell 101

Bar No.	Heat treatment	Hardness			Location
		Brinell	Rockwell	Scleroscope	
1	Fully annealed	101	..	21-22	At factory on surface
		107 end			
		105 center	54	21-22	Half-way
		107 end			through

$\frac{3}{4}$-in cube removed from end of bar for photomicrograph specimen
Physical properties :
    Elastic limit, lb. per sq. in................. 25,300
    Yield point, lb. per sq. in.................... 25,300
    Ultimate strength in tension, lb. per sq. in.... 52,400
    Percentage of reduction in area.............. 67
    Percentage of elongation in 2 in............. 41
    Ultimate shear stress from torsion
    Ultimate shear stress by die method, lb. per sq. in. 39,400
    Elastic limit in compression, lb. per sq. in...... 22,000
    Ultimate compressive strength, lb. per sq. in...... 85,350

APPENDIX NO. 2-B

MATERIAL-RECORD-SHEET SUMMARY SHOWING THOSE PROBLEMS FOR
WHICH EACH BAR WAS USED

Material identification		Problems for which material was used
Name	Bar No.	
S.A.E. 3120 steel.....................	31	c, d, e, f
S.A.E. 2345 steel.....................	29	c, d, e, f
S.A.E. 2320 steel.....................	30	c, d, e, f
S.A.E. 1035 steel.....................	28	c, d, e, f
0.15% carbon steel....................	1	b, f
0.15% carbon steel....................	2	b, c, f
0.15% carbon steel....................	3	a, b, c, f
0.15% carbon steel....................	4	b, c, f
0.15% carbon steel....................	5	a, e
0.15% carbon steel....................	6	d
1.03% carbon steel....................	32	c, d, e, f
Cast iron	1	..
Cast iron	2	d
Cast iron	3	c
Cast iron	3A	c
Cast iron	4	b
Cast iron	4A	f
Cast iron	5	c
Cast iron	6	c
Cast iron	7	a
Cast iron	7A	e
Cast iron	8	b, f
Cast iron	8A	f
Cast iron	9	c
Cast iron	9A	c
Cast iron	10	a, d
Cast iron	11	e
Brass	24	b, c, f
Brass	25	c also cutting speeds
Brass	26	d, f
Brass	27	c, d, f
Brass	33	e, f
Brass	34	e, f

APPENDIX NO. 3

TEST BARS USED FOR OBTAINING PHYSICAL PROPERTIES OF MATERIALS CUT

1 Steel tensile test specimen.
   (Standard A.S.T.M. bar, 0.505 in. diameter.)
2 Cast-iron tensile test specimen.
   (Standard A.S.T.M. bar, 0.800 in. diameter.)
3 Cast-iron compression test specimen.
   (Cylinder 1.128 in. diameter and 2.5 in. high.)
4 Steel compression test specimen.
   (⅞ in. square section and 3 in. high.)
5 Brass tensile test specimen.

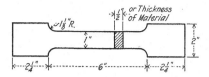

6 Shear test specimens for all metals.
     (½ in. square and 3 to 4 in. long.)

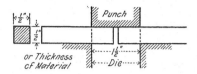

APPENDIX NO. 4

BIBLIOGRAPHY OF RESEARCH IN CUTTING TOOLS AND METAL CUTTING

GENERAL

Bibliography of the Manufacture, Heat Treatment, Uses and Tests of High Speed Steel, American Society for Steel Treating, October 1922, pp. 47-75.

A — CUTTING FLUIDS

A-1 Lubrication of Cutting Tools, E. K. Hammond.
       Machinery, Jan. 1917.

A-2 Cutting Oils, Their Properties, Composition, Examination, and Industrial Application, C. W. Copeland, Remington Arms Co.
       Chemical & Metallurgical Engineering, July 1, 1917, p. 25.

A-3 The Action of Lubricants on Metal Cutting Tools, Albert Kingsbury and others.
       Discussion of Mr. DeLeeuw's paper, A Foundation for Machine Tool Design, *Journal,* A.S.M.E., July, 1917, p. 582.

A-4 Memorandum on Cutting Fluids, Bulletin No. 2, Dept. of Scientific and Industrial Research, London, 1918.

A-5 Klean Kut Facts, and
       Up to the Minute Scientific Comments on Oil Products Used for Metal Working and Heat Treating.
       D. A. Stuart & Co., Inc., 2727 S. Troy St., Chicago.

A-6 Oil Analysis, A. H. Gill.
       J. B. Lippincott Company, 1919.

A-7 Industrial Oil Engineering, J. R. Battle.
       J. B. Lippincott Company, 1920, pp. 635-638.

A-8 Cutting Fluids, E. C. Bingham, Chemist.
       Bureau of Standards Technologic Paper No. 204, 12-20-21.

A-9 Infection from Cutting Oils, A. L. DeLeeuw.
       American Machinist, Dec. 14, 1922, p. 915.

A-10 Cutting Lubricants and Their Application; Principles of Coolants; Suitability to Material Being Cut; Factors Such as Power Consumption and Removal of Chips, A. A. Dowd and F. W. Curtis.
       American Machinist, Oct. 25, 1923, p. 613.

A-11 Use and Importance of Cutting Lubricants; Proper Application to Tool and Work; Value of Lubricants in Thread Cutting, A. A. Dowd and F. W. Curtis.
American Machinist, Jan 24, 1924, p. 135.

A-12 Lubrication and Cooling the Cutting Tool, Allan F. Brewer, Texas Oil Company.
Industrial Management, April 1924, p. 197.

A-13 Storing and Clarifying Oil in Shops.

A-14 Modern Cutting and Grinding.
Oakley Chemical Co., Thames St., New York City.

A-15 Industrial and Automotive Lubrication, Waverley Oil Works Co., Pittsburgh, Penna. (Second Edition 1925.)

A-16 Compounds for Cutting, Grinding, and Rust Prevention, J. A. Maguire.
Western Machinery World, Feb. 1926, p. 65.

A-17 Machinery's Handbook, p. 489, and p. 878.

A-18 Mark's Handbook, p. 645.

A-19 Patents on Cutting Oil, Maitland, Sun Oil Company.

B — DRILLING, TAPPING, AND REAMING

B-1 A Twist Drill Dynamometer, W. W. Bird and H. P. Fairfield.
Trans. A.S.M.E., vol. 26, 1905, p. 355.

B-1a Experiments on the Forces Acting on Twist Drills — When Operating on Cast Iron and Steel, Dempster Smith and R. Poliakoff.
Proc., Inst. of Mech. Engrs., March 1909, p. 315.

B-2 Radial Drilling Machines Break Drills and Records, F. E. Bocorselski.
American Machinist, vol. 33 (1910), p. 379.

B-3 High Speed Steel Drill Tests, F. R. Norris.
American Machinist, vol. 34 (1911), p. 719.

B-4 The Development of the Drill Test as a Means of Ascertaining the Machining Properties of Iron and Other Metals and for the Investigation of Tool Steels, A. Kessner.
Iron and Steel Inst., Carnegie Schol. Mem., vol. 5 (1913), p. 10.

B-5 An Efficiency Testing Machine for Testing Drills, Taps, and Dies, T. Y. Olsen.
Proc. A.S.T.M., vol. 14, pt. 11 (1914), p. 541.

B-6 Experiments with Drilling Dynamometer. (New drilling data from 6-ft. radials — results of Dempster Smith of England. Drills ¾-in., 1-in., to 3-in. diam. forged-twisted. H. M. Norris.
American Machinist, Aug. 13, 1914, p. 265.
Kent's Handbook, 9th ed. pp. 1286-1288.

B-7 An Investigation of Twist Drills, Bruce W. Benedict and W. Penn Lucas.
Bull. No. 103, Engrg. Exp. Sta., Univ. of Illinois, 1917.

B-8 Twist Drill Dynamometer, R. Poliakoff.
American Machinist, June 12, 1919, p. 1132.

B-9 Effect of Design on Drilling Machine Efficiency; Discusses speeds and feeds and effect of high-speed steel on design of machine and spindle of drive, F. E. Johnson.
Machinery, Aug. 1922, p. 964.

B-10 Significance of Tool Temperatures (Thermocouple in point of drill), H. H. Schwartz and W. W. Flagle.
A.S.T.M. *Journal,* June 1923.

B-11 Foreign Progress in Cutting Metals, C. A. Beckett.
Apparatus developed by Drs. Heym and Kessner to determine machinability factors affected by (a) drill hardness, (b) durability of cutting edge, and (c) variable cutting angle of drill.
Mechanical Engineering, Oct. 1924, p. 624.

C — MILLING

C-1 On the Art of Milling, John Airey and C. J. Oxford.
     Mechanical Engineering, vol. 45 (1921), p. 549.
C-1a Milling Cutters and Their Efficiency, A. L. Deleeuw.
     Trans., A.S.M.E. vol. 33 (1911), p. 245.
C-2 Effect of Variation in Design of Milling Cutters on Power Require-
     ments and Capacity, B. P. Graves and J. A. Hall.
     Mechanical Engineering, vol. 45 (1923), p. 155.
C-3 Power Required for Cutting Metals by Milling, Fred A. Parsons,
     Kempsmith Milling Machine Company.
     Mechanical Engineering, January 1923, p. 35.
C-4 Milling with Stellite Cutters.
     Machinery, April 1923, May 1923, p. 717.
C-5 Chart for Calculating S. & F. of Solid and Inserted-blade Milling
     Cutters, Frank W. Curtis, Haynes-Stellite Co.
     American Machinist, Sept. 27, 1923, p. 472.
C-6 Chip Formation by Milling Cutters (with photographs of suc-
     cessive stages rake 0 to 20 deg.), C. F. Roby (thesis at
     Cincinnati Univ.).
     Mechanical Engineering, Oct. 1923, p. 586.
C-7 A Treatise on Milling and Milling Machines, Cincinnati Milling
     Machine Co., 1922.

D — TURNING, STRAIGHT LINE, AND GENERAL

D-1 Cutting Tools, Robert H. Smith.
     Cassell & Co., 1881. (See Proc., Inst. of Mech. Engrs. 1903,
     p. 48.)
D-2 A Novel Method of Testing C. I. for Hardness, Chas. A. Bauer.
     American Machinist, 1-4, 1897, S-245.
D-3 Schnelldrehstahl — Bericht des vom Berliner Bezirksverin
     deutcher Ingenieure Gebildeten Ausschusses.
     Zeitschrift des Vereines deutscher Ingenieure, vol. 45 (1901),
     p. 1377.
D-4 Experiences Sur Le Travail Des Machines — Outils, Pour Les
     Metaux, C. Codron.
     Vol. I (1901-1902) and vol. II (1903-1905). Extrait du *Bulle-
     tin de la Societe d' Encouragement pour l' industrie Nation-
     ale*. Vol. III (1906-1910), Extrait de la *Revue de Mecanique*.
     Reprints: vol. I (1902) considers drilling, vol. II (1906)
     considers reaming, vol. III (1910) considers general machine
     tools, shearing, and punching. G. E. Stechert & Co., 33 E.
     10th Ave., New York, N. Y.
D-5 Notes on High Speed Steels (as used in Union Pacific Railroad
     Shops at Omaha, Nebr.), Henry H. Suplee.
     Proc. of Inst. of Mech. Engrs., July 1903, p. 457.
D-6 Cutting Angles of Tools for Metalwork, as Affecting Speed and
     Feed (lathe dynamometer used), H. F. Donaldson.
     Proc. of Inst. of Mech. Engrs., Jan. 3, 1903, p. 5.
D-7 Experiments with a Lathe Tool Dynamometer, J. T. Nicolson.
     Trans. A.S.M.E., vol. 25 (1903), p. 637.
D-8 On the Art of Cutting Metals, F. W. Taylor.
     Trans., A.S.M.E., vol. 28 (1907), p. 31.
D-8a Discussions by H. R. Towne, J. M. Dodge, C. W. Rice, J. S.
     Bancroft, H. K. Hathaway, J. T. Hawkins, L. P. Brecken-
     ridge.
     Proc. A.S.M.E., Jan. 8, 1907, p. 929.
D-8b Discussions by J. E. Stead, Oberlin Smith, W. H. Blawelt, D.
     Adamson, and W. S. Huson.
     Proc., A.S.M.E., Mar. 1907, p. 1159.

D-8c Discussions by H. LeChatelier, C. Codron, W. Lewis, R. A. Skeggs
J. T. Nicolson, F. W. Taylor.
Proc., A.S.M.E., April 1907, p. 1339.

D-9 The Cutting Properties of Tool Steels, Edward G. Herbert.
Jour. Iron & Steel Inst., 1910, pt. 1, p. 206.

D-10 Power Required to Remove Metal, Chas. Robbins.
Trans., A.S.M.E., vol. 32 (1910), p. 202.

D-11 High Speed Steel, O. M. Becker.
McGraw-Hill, 1910.

D-12 Boring Tools, Dempster Smith.
Trans. Manchester Assn. of Engrs., Oct. 1911.

D-13 Cutting Force, Power, and Drive in Lathe Work, G. Lindner.
Zeits. f. Werkzeugmaschinen u. Werkzeuge, Feb. 5, 1912.

D-14 The Influence of Heat on Hardened Tool Steels, Edward G. Herbert.
Jour. Iron & Steel Inst., vol. 85 (1912 pt. 1), p. 358.

D-15 Tool Steel for the U. S. Navy, Lewis H. Kenney.
Society of Naval Arch. & Marine Eng., 1912.

D-16 Cutting Power of Lathe Turning Tools, William Ripper and C. W. Burley.
Proc., Inst. of Mech. Engrs., 1913, p. 1067. For further discussion, see *Prac. Engr.*, Dec. 11, 1913 and later (serial).

D-17 Die Fortschritte deutscher Stahlwerke. (The Progress of German Steel Industry). *Stahl und Eisen*, 1913.

D-18 Power Requirements of Lathes, A. Bodi.
Industria, Sept. 14, 1913.

D-19 A Basis of Measuring Lathe Capacity, L. R. Pomeroy.
American Engineer, March 1913 (discusses Nicolson's results).

D-20 Past Experimental Work on Cutting Tools, Dempster Smith.
Proc., Manchester Assn. Engrs., March 1915.

D-21 Notes on the Relations Between the Cutting Efficiencies of Tool Steels and Their Brinell and Scleroscope Hardness, J. O. Arnold.
Jour., Iron & Steel Inst., vol. 93 (1916 pt. 1), p. 102.

D-22 A Foundation for Machine-Tool Design and Construction, A. L. DeLeeuw.
Trans., A.S.M.E., vol. 39 (1917), p. 185.

D-23 Molecular Constituents of High Speed Steels and Their Correlation with Lathe Efficiencies, J. O. Arnold and Fred Ibbotson.
Jour. Iron & Steel Inst., vol. 99 (1919, pt. 1), p. 407.

D-24 Die Werkzeugmaschinen, 3 Aufl., Hulle.
(Machine Tools — 3 Editions) Springer, Berlin, 1919.

D-25 Der Schnellbetrieb (High Speed Work), L. Walther.
R. Oldenbourg, 1919, Munchen.

D-26 Die Dreherei und ihre Werkzeuge (Turning and its Tools), 3 ed., Willie Hippler.
Springer, Berlin, 1919.

D-27 Cutting Edges, R. E. Crompton.
Engineer, vol. 130 (1920), p. 207.

D-28 Uber Dreharbeit und Werkzeugstahle (About Turning and Cutting Tools), Taylor-Wallichs.
Julius Springer, 1920.

D-29 Comparative Tests upon High Speed Steels, A. J. Langhammer.
Chemical & Metallurgical Engineering, vol. 22 (1920), p. 889.

D-30 Cutting Pressure and Form of Chip, H. Sack.
Betrieb, vol. 3, Sept. 15, 1921.

D-31 Hardening High Speed Steel, A. H. d'Arcambal.
Chemical & Metallurgical Engineering, vol. 25 (1921), p. 1168.

92   RESEARCH IN THE ELEMENTS OF METAL CUTTING

D-32 Metal Cutting Tools, Their Principles, Action, and Construction,
      A. L. DeLeeuw.
      McGraw-Hill, 1922.
D-33 Report of Lathe Tools Research Committee, Dempster Smith.
      Bul. Manchester Assn. of Engrs. 1922.
D-34 Lathe Breakdown Tests of Some Modern High Speed Steels,
      H. J. French and Jerome Strauss.
      Trans., A.S.S.T., vol. 2 (1922), p. 1125.
D-35 An Account of Some Experiments on the Action of Cutting Tools,
      E. G. Coker and K. C. Chakko.
      Excerpt Minutes of Proceedings of Meetings, Bul. pub. by
      Inst. of Mech. Engrs., April 1922.
D-36 Are Metal Workers Hidebound?
      Editorial, *Amer. Mach.*, 1922, vol. 57, p. 154.
      Discussion by Alfred Herbert, Ltd., *American Machinist*, 1922,
      vol. 57, p. 768.
      Discussion by F. W. Curtis, *American Machinist*, vol. 58, 1923,
      p. 162.
D-37 Considerations Generales sur Les Machines-outile. *Outillage*,
      vol. 6, nos. 37, 38, 39, 43, 44, 45, 47, Sept. to Nov. 1922.
      (Power required, theory of design, etc.)
D-38 Causes of Failure of High Speed Tools, J. F. Kayser.
      Chemical & Metallurgical Engineering, vol. 25 (1922), p. 847.
D-39 Performance of High Speed Cutting Tools, Jerome Strauss.
      Iron Age, vol. III (1923), p. 1103.
D-40 Power Required for Cutting Metal, Fred A. Parsons, Kempsmith
      Milling Machine Co.
      Mechanical Engineering, vol. 45 (1923), p. 35.
D-41 Significance of Tool Temperatures as a Function of the Cutting
      Resistance of Metals, H. A. Schwartz and W. W. Flagle.
      Proc. A.S.T.M., vol. 23, pt. 11 (1923), p. 27.
D-42 Effect of Heat Treatment on Lathe Tool Performance and Some
      Other Properties of High Speed Steels, H. J. French, Jerome
      Strauss, and T. C. Digges.
      Trans. A.S.S.T., vol. 4 (1923), p. 353.
D-43 Engineering Problems Solved by Photo-Elastic Methods, E. G.
      Coker.
      Jour. Franklin Inst., 1923, vol. 196, p. 433.
D-44 Metal Cutting Tools (discussion of tools shapes, angles, and
      speeds), W. E. Splain.
      American Machinist, p. 663, vol. 58, May 3, 1923.
      Discussion by F. W. Elstub, *American Machinist*, p. 64, vol. 59,
      July 12, 1923.
      Discussion by R. Poliakoff, *American Machinist*, p. 783, vol. 59,
      Nov. 15, 1923.
D-45 Practical Notes on Speeds and Feeds, Giving Tables of Speeds
      for Different Processes and Metals.
      American Machinist, Sept. 27, 1923, p. 467.
D-46 Metallurgical Aspect of Cutting Metals.
      Iron Age, Sept. 27, 1923, p. 805.
D-47 Lathe Tool Angles and Speeds for Cutting Monel Metal and
      Nickel.
      Machinery, Dec. 1923, p. 287.
      Also see International Nickel Co.'s Working of Monel.
D-48 Carbon Tool Steel Specifications, A.S.T.M.
      American Machinist, Dec. 20, 1923, p. 914.
D-49 Results of Extended Work on Tool Form, Force, Chip Formation,
      Speed, Etc., Hans Klopstock.
      Werkstättstechnik, Dec. 15, 1923, Part 1. Berlin, Part 2.

D-50 Progress Report on the Present Status and Future Problems on the Art of Cutting Metals, C. A. Becket.
A.S.M.E. Special Research Committee, A.S.M.E. Dec. 3, 1923.

D-51 Investigation of Turning, Hans Klopstock.
Werkstättstechnik, vol. 19, Dec. 1 and 15, 1923.

D-52 Tacho-Scope, Direct Reading R.P.M. Counter and F.P.M.
American Machinist, vol. 59, p. 155.

D-53 Schnellstahl und Schneidmetall (High Speed Steel and Tool Steel), F. Rapatz.
Maschinenbau, 1924.

D-54 Roto-stat, A Means of Slowing up Motion of Cutting so as to Make, it Readable.
American Machinist, vol. 59, p. 349, and Jan. 17, 1924, p. 103.

D-55 Speeds and Feeds for Materials, Effect of Lubrication, Etc., Dowd and Curtis.
American Machinist, Feb. 1924.

D-56 Zur Normung der Schneidstahle Ruckblick und Ausblick (Standardization of Cutting Tools, Terms, etc.).
N.D.I., Feb. 28, 1924, p. N-63-65.

D-57 Influence of Temperature on Work-Hardening of Metals, E. G. Herbert.
Engineer, April 4, 1924, p. 356.
Also see p. 366 Smith & Hey's article.

D-58 Hardness and Cutting Trials of Tool Steel, Dempster Smith and I. Hey.
(Manchester Assn. of Engrs, March 28, 1924.)
Engineer, April 4, 1924, p. 363.

D-59 The Thermal Action of Cutting Tools, Dempster Smith and I. Hey.
Editorial, April 4, 1924, *Engineer*, p. 363.

D-60 Testing Machinability by Drilling, Kessner.
Testing, April, 1924.

D-61 Chip Formation, as Tested at National Physical Laboratory, England.
Engineering, May 16, 1924.

D-62 Definition of Terms Used in Heat Treatment.
Tentatively approved by Recom. Prac. Comm. of A.S.S.T.
American Machinist, Mar. 20, 1924, p. 420.

D-63 Foreign Progress in Cutting Metals, C. A. Beckett.
Mechanical Engineering, Oct. 1924, Progress report of Special Research Committee, covering work of Smith and Hey, and Herbert hardness tester.

D-64 Schneidversuche mit Schnellarbeitsstahlen, R. Hohage and A. Grutzner.
Stahl und Eisen, vol. 45, 1925, p. 1126.

D-65 Report on the Action of Cutting Tools, E. G. Coker.
(Use of polarized light on lathe, planer, and milling tools.)
Engineering, vol. 119 (1925), p. 365.
I.M.E. paper.

D-66 Experimental Study of Forces Exerted upon Surfaces of a Cutting Tool, T. E. Stanton and Mr. J. H. Hyde of National Physical Laboratory, London.
Engineer, Jan. 30, 1925, p. 126.
Engineering, Jan. 30, 1925, p..139, 148, and 151.

D-67 Flow and Rupture of Metals during Cutting, W. Rosenhain and Mr. A. C. Sturney of National Physical Laboratory, London.
Engineering, Jan. 30, 1925, p. 151.
Engineer, Jan. 30, 1925, p. 126.

D-68 Thermoelectric Measurement of Cutting Tool Temperatures, Henry Shore.
Jour., Washington Academy of Sciences, March 4, 1925, p. 85.

D-69 Factors Influencing the Machinability of Cast Iron, John W. Bolton, Metallurgist, Niles Tool Wks.
Machinery, March 1925, p. 533.

D-70 Der heutige Stand der Zerspanungs wissenschaft (The Present State of Metal Cutting), W. Hippler.
Zeitschrift für Maschinenbau, 1925.

D-71 Experiments with Lathe Tools on Fine Cuts and Some Physical Properties of the Tool Steels and the Metal Operated Upon, Dempster Smith and Arthur Leigh.
Engineering, March 20, 1925, p. 364.
Engineer, March 30, 1925, p. 319.
Paper of Institution of Mech. Engr.

D-72 Recent Investigations in Turning and Planing and a New Form of Cutting Tool, Hans Klopstock.
Mechanical Engineering, June 1925.
Zeits. des Vereines deutcher Ingenieure, Feb. 21, 1925.

D-73 A Lathe Dynamometer Described, A. Sahnazaroff.
Die Werkzeugmaschine, May 15, 1925, p. 254.

D-74 Tools, Speeds, Feeds, and Depth of Cut for Finishing Commutators (Copper and Micaplate).
Machinery, June 1925, p. 817.

D-75 Wissenschaftliche Gestaltung der Werkzeuge (Scientific forming of Cutting Tools).
Zeitschrift des Vereines deutcher Ingenieure, 1925.

D-76 Spanalhebende Werkzeuge fur die Metallbearbeitung und ihre Hilfsernrichtungen (Chip taking tools for the Working of Metals and their Accessories), J. Reindl.
Springer, Berlin, 1925.

D-77 Cutting Speeds of Crank Shapers.
Machinery, Aug. 1925, p. 957.

D-78 A New Method of Measuring Temperatures Generated by Metal Cutting Tools, Edward G. Herbert.
(Thermoelectric, stellite tool cutting mild steel.) Description, *The Pendulum*, June 1925.
Some data, *The Pendulum*, Sept. 1925.

D-79 Experiments with Nickel, Tantalum, Cobalt, and Molybdenum in High Speed Steel, H. J. French and T. C. Digges.
Annual Meeting A.S.S.T., Sept. 1925.

D-80 Die Prufung der Gewindebohres (The Investigation of Taps, at Charlottenburg Laboratory for Machine Tools "Torque"), Von M. Kurrein.
Werkstattstechnik, Sept. 1, 1925.

D-81 Hardness Governs Machinability, James Ward.
Foundry, Sept. 1, 1925, p. 685.

D-82 Why Do We Know so Little About Metal Cutting?
American Machinist, Sept. 17, 1925, p. 476.

D-83 Machinability of Brass Rod in Screw Machines, A. C. Lusher, Scoville Mfg. Co., Providence, R. I.
Paper before Providence Section, A.S.M.E., Oct. 1925.

D-84 Hardness Numbers and Their Relation to Machinability.
Mechanical Engineering, Nov. 1925, p. 919.
Discussion at Cleveland Meeting of A.S.S.T. in Sept. 1925.

D-85 Research in Metal Cutting, O. W. Boston.
American Machinist, Nov. 19, 1925, p. 805.

D-86 Physical Metallography, E. Heyn. Translated into English by M. A. Grossmann.
(Excellent bibliographical index.)
(Relation of Machinability to Hardness and Workability), p. 326.
John Wiley & Sons, 1925 — United Alloy Steel Corp.

D-87 Die Messung der Schneidentemperature beim Abdrehen von Flusseisen (Measuring the cutting temperatures in the cutting of low carbon steel), Gottwein.
Maschinenbau, Nov. 19, 1925, p. 1129-1135.

D-88 The Measurement of Cutting Temperatures, E. G. Herbert.
Paper, Institution of Mech. Eng., Feb. 19, 1926.

D-89 Cutting Tests of Tool Steels, U. S. Naval Gun Factory, Jerome Strauss.
Trans., A.S.S.T., April 1926.

D-90 Supplementary paper on The Measurement of Cutting Temperatures, E. G. Herbert.
American Machinist, March 4, 1926, p. 363.

D-91 Cutting, Principles of Cutting Edges, W. R. Ward.
Trans., A.S.S.T., March 1926, p. 482.